Weather and Health

Born in Frankfurt, Germany, in 1906, HELMUT E. LANDSBERG studied at the University of Frankfurt and received his Ph.D. in geophysics and meteorology there. He then joined the faculty of Pennsylvania State University where he initiated courses in meteorology and geophysics and conducted research in the field of air pollution. After serving on the faculty of the University of Chicago, he worked as consultant and operations analyst for the Air Force during World War II.

In 1954, Dr. Landsberg became Director of Climatology of the U. S. Weather Bureau, and when the Environmental Science Services Administration was formed, he became the first director of the Environmental Data Service.

Dr. Landsberg has been elected to membership in the National Academy of Engineering, and to fellowship in the American Academy of Arts and Sciences, the American Geophysical Union (of which he is currently President), the American Meteorological Society, the Meteoritical Society, and the American Association for the Advancement of Science. He is also serving as President of the American Institute of Medical Climatology.

In 1966, Dr. Landsberg resumed his academic role and is now research professor and Chairman of the Graduate Committee on Meteorology at the University of Maryland.

WEATHER
and
HEALTH

An Introduction to Biometeorology

HELMUT E. LANDSBERG

1969
DOUBLEDAY & COMPANY, INC.
GARDEN CITY, NEW YORK

SELECTED TOPICS
IN THE ATMOSPHERIC SCIENCES

The American Meteorological Society, with the objectives of disseminating knowledge of meteorology and advancing professional ideals, has sponsored a number of educational programs designed to stimulate interest in the atmospheric sciences. One such program, supported by the National Science Foundation, involves the development of a series of monographs for secondary school students and laymen, and since the intended audiences and the standards of excellence were similar, arrangements were made to include their volumes on meteorology in the Science Study Series.

This series within a series is guided by a Board of Editors consisting of James M. Austin, Massachusetts Institute of Technology; Richard A. Craig, Florida State University; James G. Edinger, University of California, Los Angeles; and Verne N. Rockcastle, Cornell University. The society solicits manuscripts on various topics in the atmospheric sciences by distinguished scientists and educators.

THE SCIENCE STUDY SERIES

The Science Study Series offers to students and to the general public the writing of distinguished authors on the most stirring and fundamental topics of science, from the smallest known particles to the whole universe. Some of the books tell of the role of science in the world of man, his technology and civilization. Others are biographical in nature, telling the fascinating stories of the great discoverers and their discoveries. All the authors have been selected both for expertness in the fields they discuss and for ability to communicate their special knowledge and their own views in an interesting way. The primary purpose of these books is to provide a survey within the grasp of the young student or the layman. Many of the books, it is hoped, will encourage the reader to make his own investigations of natural phenomena.

The Series, which now offers topics in all the sciences and their applications, had its beginning in a project to revise the secondary schools' physics curriculum. At the Massachusetts Institute of Technology during 1956 a group of physicists, high school teachers, journalists, apparatus designers, film producers, and other specialists organized the Physical Science Study Committee, now operating as part of Educational Development Center, Newton, Massachusetts. They pooled their knowledge and experience toward the design and creation of aids to the learning of physics. Initially their effort was supported by the National Science Foundation, which has continued to aid the program. The Ford Foundation, the Fund for the Advancement of Education, and the Alfred P. Sloan Foundation have also given support. The Committee has created a textbook, an extensive film series, a laboratory guide, specially designed apparatus, and a teachers' source book.

Contents

ONE

The Essential Atmosphere

Air is essential to human survival. Even when man ventures into outer space he has to take a bit of atmosphere along. On earth the atmosphere is that wild ocean of air in continuous flux, driven by solar energy, at the bottom of which we live. Its behavior and motion are the subject of meteorology, the science of the atmosphere.

The phase of the science dealing with the relations between atmospheric and life processes is called biometeorology. It is a vast field. Forests and crops depend on the weather. Animals mate, migrate, and hibernate in annual cycles, dictated by the course of weather through the year. And much as we humans like to think that we have gained independence from natural forces, we too are greatly governed by the weather, in our activities and in the reactions of our bodies. Meteorologists try, of course, to give ample warning about weather hazards, such as tornadoes, hurricanes, and blizzards, but this is not our concern here. We want rather to take a look at the more subtle influences of weather on human well-being, the little sneak attacks of the environment on us. We are often quite unaware of them until we sweat profusely, or have a sunburn or frostbite, or even a spell of arthritic pain.

Although the relations of weather and climate to health and comfort have been studied since antiquity there are still many unsolved problems. The reason for the remaining puzzles has to be sought in the complexity of both, the atmospheric events and the intricate functioning of the human body. Weather is proverbially fickle and each change in the weather requires an adaptation of the human body. Much of that occurs quite unnoticed by us. The physiological mechanisms are very efficient—up to a point. Yet, sometimes they break down and then trouble starts, a pathological reaction takes place that may result in injury, illness, or even death. The weather also affects our moods and steers some of our psychological responses.

Biometeorologists have been trying for several decades to unravel these relationships. For some of the simpler ones they have succeeded. For the more complicated ties the answers are still missing. These more direct and readily understood influences of the atmosphere on human well-being will be discussed first. The more subtle actions of the atmosphere lead us to the present frontier of knowledge, to which the later chapters will be devoted.

TWO

Air and Altitude

Man can survive without food for weeks. He can get along without water for a few days. But without air he will suffocate in a few minutes. Just what is this precious air?

It is essentially a mixture of gases. In the present geological era its primary constituent is nitrogen (N). There is 78 per cent N, by volume, in air. Next comes oxygen (O_2) with 21 per cent, the unique link between the atmosphere and the life process. The total of minor constituents in the air accounts for about 1 per cent. Among them are carbon dioxide (CO_2) and ozone (O_3), also important in life processes. Carbon dioxide is utilized in plant photosynthesis during which it helps to form organic substances. In this process the plants give off oxygen. It is now generally accepted that plants were on earth earlier than animals and that their activity brought the oxygen into the atmosphere, paving the way for oxygen-breathing animal life and leading ultimately to the human life. The ozone, a triatomic oxygen, although present only in minute quantities, also plays an important part in life processes. It originates in the high atmosphere by a photochemical process, caused by sunlight of short wave length. The highest ozone concentration is at 25 kilometers, but occasionally ozone forms also near the earth's surface; this happens, for example, when the atmosphere is highly polluted by car exhaust. Ozone is irritating to mucous membranes, and is one of the offenders in smog.

At the earth's surface the amount of oxygen in the air stays essentially constant irrespective of the changing weather or seasonal climates. But its absolute amount changes with elevation above sea level. The biometeorological effects of change in elevation are caused by the decreased density of the air and the resulting decrease in the partial pressure of oxygen. While the average pressure at sea level is about 1013 millibars, the pressure is about halved at 5.5 kilometers, and cut to one-quarter at 11 kilometers. This is somewhat higher than Mount Everest (8840 meters), the highest point on earth. At this level the oxygen is insufficient for survival. If man goes higher in a spacecraft, he needs the help of an artificial atmosphere.

The problem was encountered early in the history of flight. In fact, it was first noted on a flight made for meteorological exploration by James Glaisher, Superintendent of Meteorological Observations at the Greenwich Observatory in England. The year was 1862, but even though manned balloon flight was nearly eighty years old, that year marked the beginning of scientific exploration of the free atmosphere. Glaisher wanted to learn more about upper winds, clouds, and temperatures at higher altitude. After several ascents in July and August 1862, his balloonist Henry Coxwell attempted a record flight on September 5. They claimed to have reached a record height of 11.3 kilometers. Yet Glaisher had been troubled by blurred vision and passed out completely at about 5.8 kilometers. Coxwell could barely pull his ripcord to bring the balloon down. But even 6 kilometers was a respectable height.

Three French aeronauts who tried for an altitude record over a decade later, on April 15, 1875, ran into serious trouble. Their balloon was fittingly named *Zenith* and equipped with primitive oxygen apparatus. Gaston Tissandier and his companions, Croci-Spinelli and Sivel, did not use the oxygen properly, and all passed out at 7.6 kilometers. Only Tissandier survived.

These explorations led to scientific investigations of the effects of oxygen deficiency, called hypoxia.

The gas mixture of the air with 78 per cent nitrogen, 21 per cent oxygen, and 1 per cent of minor constituents stays remarkably constant to about 70 kilometers height. The pressure there is only around .05 millibar or .005 of 1 per cent of the surface pressure. The constancy of mixture means that the pressure of oxygen decreases with height at the same rate as that of the air as a whole. The partial pressure of oxygen at sea level is about 212 millibars. At 3 kilometers its partial pressure is reduced to 146 millibars, and at 10 kilometers only about 55 millibars oxygen pressure prevails.

Oxygen is needed for the body's metabolism. This need does not change with altitude above sea level. For proper functioning and survival, the body needs as much oxygen at higher levels as at sea level. This poses a severe problem when a person is rapidly transferred from a lower to a higher altitude. A change from sea level to 2 kilometers elevation is generally without consequence. The body can evidently cope with the approximately 20 per cent reduction in partial pressure of oxygen, but at higher levels oxygen deficiencies begin to show. For this reason, modern, high-flying airplanes are artificially pressurized to keep oxygen pressures at least comparable to the 2 kilometer level.

At moderate elevations above that critical level, hypoxia leads to headache, tiredness, and occasionally to nausea. At higher elevations an individual adapted to sea level will start breathing faster, and his heart rate will speed up. If his rapid breathing continues, the loss of carbon dioxide from the blood may become too large. The chemical composition of the blood is altered and circulatory difficulties to the verge of collapse may be the result.

Above 6 kilometers another critical limit is reached after short exposures. The brain becomes seriously affected by lack of oxygen. At first simple mental tasks become difficult, writing is incoherent, and some persons lose consciousness altogether, as happened on Glaisher and Coxwell's early high-altitude balloon flight. At present, all high-altitude airplanes carry oxygen for emergencies that may cause lessening of pressure. The crews breathe oxygen at all times above certain levels to assure safe operation. Short exposures to hypoxia are readily reversible.

People who are called upon to perform work at unaccustomed elevations need fairly extended periods of acclimatization. The organism develops deeper breathing, perhaps even an increase in lung capacity. Next comes an increase in the number of red blood corpuscles. These are the oxygen carriers of the blood. If more of them are present, more oxygen is absorbed in each breath and carried to muscles and tissues needing the life-sustaining gas. In mountain climbing it has been necessary for expeditions to carry through their attempts on high peaks in stages. Exceptionally well trained and acclimated climbers, sometimes supported by auxiliary oxygen, have been able to reach even the highest peaks. Yet the exceptional altitudes are not places where humans can survive for any length of time. There are, of course, native populations at considerable heights. In the United States many persons live in the "mile-high" plateau east of the Rockies without apparent impairment. Mining operations are regularly carried on quite normally at 3 kilometers in the Rocky Mountains. At these heights, where there is only 70 per cent of the oxygen pressure at sea level, healthy acclimated persons perform as well as their counterparts in the lowlands.

In fact, there are two major cities in the world, La Paz, Bolivia (3600 meters), and Lhasa, Tibet (3900 meters), where life perks along merrily with only about 62 per cent of the oxygen present at sea level. Indeed, South American miners work normal shifts at 5800 meters in the high Andes and after work may engage in a strenuous soccer game without ill effects. They have less than half of the sea-level oxygen available. Nature, however, has endowed these natives with much larger lung capacity than their sea-level compatriots. They are barrel-chested and the total surface of lung tissue, where the oxygen–carbon dioxide exchange takes place, is vastly increased. The number of red blood corpuscles, which at sea level runs around 4.75 million per cubic millimeter, is over 8 million per cubic millimeter in the mountain Indians of the Andes. Yet these strong men of the mountains are miserable if they are transferred to sea level. Even after a prolonged stay some never completely live in harmony with the environment. The young and healthy of all races have the physiological elasticity to acclimatize in a few weeks or months to elevation changes of 3 or even 4 kilometers, up or down. For older or infirm persons such changes in environment are detrimental or even dangerous. It is fortunate that only a fraction of 1 per cent of the earth's surface is above these elevations so that with the few exceptions of mining mineral deposits, man's presence there is for sport or in search of the mythical abominable snow men rather than for important daily needs.

THREE

Sun and Sunburn

The sun's radiation is a weather element of vital importance, furnishing the primary energy that drives the atmosphere. The difference in heat received between pole and low latitudes causes temperature differences which bring into action a whole host of energy changes that cause wind and weather. We are primarily concerned here, however, with the sun's rays that pass through the air and impinge upon the body.

From meteorological observatories we can readily obtain data on duration of sunshine and amount of cloudiness. Many stations also record the total incoming radiation on a horizontal surface. This is called the global radiation. By simple geometrical calculations it can be related to a vertical house wall, a slanted roof, or a person standing upright. In meteorology, radiation is usually expressed in calories per square centimeter per minute (cal cm^{-2} min^{-1}). The calorie is a very convenient heat unit in this connection because it will crop up again as a basic quantity in nutrition and heat production of the body.

For biological purposes, it is convenient to divide the total radiation in the atmosphere into various spectral bands. The solar radiation reaching the earth covers a wide span of wave lengths in the spectrum of electromagnetic waves. Figure 1 shows the various ranges. In the middle of the range are the light waves, to which our eyes

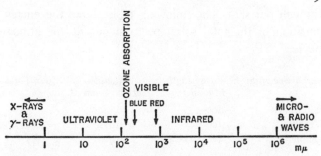

FIGURE 1. *The major portions of the electromagnetic wave spectrum (on logarithmic scale). The extra-atmospheric radiation of the sun ranges over a broad part of this spectrum. The visible comprises only a small region. Upper-atmospheric ozone absorbs most of the ultraviolet before it can penetrate to the surface.*

are sensitive. They cover the spectrum from about 400 millimicrons (violet and blue light) to 750 millimicrons (red). In the middle are the green-yellow waves, which carry considerable energy. The mixture of all the waves appears to us as white. The shortest light waves are considerably scattered by air molecules, making the sky appear blue. Beyond the shorter end of the visible waves is the ultraviolet region. These rays are partly absorbed in the high atmosphere where they create the ozone layer. It forms a barrier for all of the shortest ultraviolet wave lengths, which is fortunate because these rays are damaging or even lethal to living organisms. Some of the longer ultraviolet rays, however, penetrate through the atmosphere, especially if the air is clean. This ultraviolet radiation is of great biological importance.

Beyond the longer end of the visible spectrum lies the infrared region, which includes about half of the solar radiation. This is the dark heat radiation, sensed only through our skin. The following table shows the energy partition of the total solar spectrum outside the atmosphere.

Part of Spectrum	Wave Lengths Millimicrons	Intensity Cal cm.$^{-2}$ min^{-1}	% of Total
Ultraviolet	200–400	0.140	7
Visible	400–750	0.910	46
Infrared	750–24,000	0.930	47

Practically no ultraviolet radiation below 350 millimicrons penetrates to the earth's surface at sea level. The total solar and sky radiation there is rarely more than 75 per cent of the energy received outside the atmosphere.

The dark heat radiation emanates not only from the sun, but from all bodies having a temperature above absolute zero ($-273°$ C). The human body also radiates heat from its skin. So do walls, the soil, and even the night sky. All objects exchange radiative heat. If the surface temperature of an environmental surface is higher than the human skin temperature, we receive heat from it; if it is lower, we lose heat to it. That's why the dispensers of steam heat from a furnace are called radiators.

We can make short shrift of the visible part of the solar spectrum here. Suffice it to say that the human eye has its highest sensitivity at the wave lengths around 500 millimicrons. This is approximately in the same wave band in which the solar energy is highest. Is this coincidence or evolutionary adaptation? We do not know. The fact remains, however, that midday solar intensity in most latitudes is far too bright for a direct look. This can be a dangerous, even blinding experiment. It suggests that man was adapted to the much dimmer and filtered light of the forest and only later emerged into regions where he was fully exposed to the high light intensities of an open landscape. In fact, the Eskimos who hunt in bright sunlight with added glare from reflecting snow and ice surfaces, are the inventors of highly efficient slit goggles which let only a small amount of light fall on the eye.

The infrared part of the spectrum will also get only a quick glance at this juncture. There will be more to say about this invisible heat radiation in the later discussion of the effects of heat on the body. It's enough to say here that the human skin absorbs this radiation readily. In an upright position we receive more when the sun is not much above the horizon than when it is overhead. This may account for the pleasant experience of having the sun play on our bodies in the early morning or late afternoon hours, when the air temperature may be fairly low. Lying down, of course, more radiation is received with the high solar elevations around midday. If direct solar heating is added to high temperatures, the result may not be at all pleasant. Yet in itself, infrared radiation has no harmful effects and often is the cause for feelings of comfort and pleasure in being out in the sun.

This is more than can be said about the ultraviolet portion of the solar spectrum. The ultraviolet part of the solar radiation is responsible for many photochemical reactions; in other words, it is capable of changing substances or causing them to combine. The shortest wave lengths can kill microorganisms outright. Fortunately for the survival of presently living organisms, most of the very short wave lengths below 300 millimicrons are absorbed by the ozone layer in the high atmosphere. Much of the longer ultraviolet is also very much weakened in the atmosphere. This so-called attenuation depends on the thickness of air the rays have to penetrate and its degree of cleanliness.

At low elevations of the sun above the horizon the slant rays pierce a thickness of ozone and air several times greater than when the sun is vertically overhead. Let this latter thickness be equal to 1; then the thickness at solar elevation of 40 degrees above the horizon is 1.5 times more. At 30 degrees the rays have to penetrate a layer 2 times; at 19 degrees, 3 times; at 14 degrees, 4 times as thick as when the sun is in the zenith. When it is just above the horizon the rays have to penetrate 27 times the atmospheric thickness than from overhead. Of course, at higher elevation the atmosphere becomes thinner and at about 5500 meters about half of it is below you. This level is reached only in the highest mountains. But with every gain in elevation above sea level a certain amount of protection against damaging radiation is lost.

Another important factor in penetration of ultraviolet radiation is the transparency of the air. Air that comes fresh from the polar regions is clean and clear. It will absorb little of the radiation that has passed through the ozone layer. Air that has lolled over the continents, especially over inhabited regions, becomes very turbid and filters out most of the shorter wave lengths that came through the ozone barrier. In the United States air that has stayed over the continent for a while depletes the ultraviolet radiation greatly.

The shortest wave lengths that occasionally reach the surface have one beneficial effect. They activate a substance in the skin which bears the formidable name 7-dehydrocholesterol, also called provitamin D. Radiation of less than 320 millimicrons converts this substance into vitamin D. This helpful chemical prevents rickets, and is therefore essential for health, especially in growing children, whose bones are still forming. But, as so often in nature, the evil and the good come in the same package. The identical rays that activate vitamin D in the skin also cause painful sunburn.

At this point it might be helpful to take a quick look at the structure of the human skin. On the outside is a thin layer of horny cells, called the corneum. It is about 30 microns thick. Below that is another very thin stratum, about 50 microns thick, the Malpighian layer. These two layers together form the epidermis, or outer skin. Below that is the dermis or main skin, which is about 2 to 3 millimeters thick. It has small blood vessels and nerves embedded in it, which end just below the Malpighian layer. The ultraviolet radiation penetrates into the dermis, where it causes a photochemical reaction. Although the process is not completely known, many believe that an amino acid is transformed into a substance dilating the blood vessels. This noxious substance causes reddening of the skin or even blisters. The burn may be so severe that some part of the skin is destroyed and peels. The wave lengths primarily responsible for sunburn are below 320 millimicrons. Another chemical reaction takes place with wave lengths between 330 and 400 millimicrons. These transform another amino acid, tyrosine, through a rather complicated reaction into melanin, which settles in the Malpighian layer. Melanin is a brown pigment and its appearance causes tanning. The pigmentation, whether newly formed or naturally present, offers protection from the shorter wave lengths. They do not penetrate much below the Malpighian layer in dark-skinned persons.

On days with a clean air and the sun at a high elevation it does not take long to acquire a sunburn. The skin will become red in a few minutes. For someone who has not been exposed to sunlight for a while it may only take half an hour of exposure to get a severe sunburn. Everyone knows that in spring and early summer people who have been indoors or in smoky cities for months acquire a sunburn during their first outing or visit to the beach.

Other circumstances may aggravate the danger. One of them is reflection of radiation from below. A snow cover is an almost perfect reflector and many a skier owes his bad facial sunburn to a combination of solar radiation and back radiation from snow. A sand surface can act similarly. Clouds, too, can add to the radiative load on the body. A bright sunny day with brilliant, puffy cumulus clouds will bring about a higher combined radiation load than a sun in a completely cloud-free sky.

Elevation above sea level, however, is perhaps most important in controlling radiation intensity. Most earthborn dust and pollution stays in the lower layers of the atmosphere. Thus the air at higher levels becomes progressively clearer, cleaner, and thinner, causing an increase of the relative share of the ultraviolet radiation in the total spectrum. For example, at 3 kilometers height, with the sun 50 degrees above the horizon, the ultraviolet radiation received in the mountains on a clear day is about twice that received at sea level. This means the time for getting a sunburn, even in the absence of a snow cover, is halved. Hence, mountain climbers and populations living at higher elevations have usually deeply tanned skins.

Prolonged exposure to sunlight also evokes a protective reaction. The horny layer of the skin thickens, decreasing the amount of radiation that penetrates to the deeper layers. Thus less sunburn results. In some persons the thickening of the skin will also lead to wrinkling and give the impression of aging. If this were the only end-result, one might dismiss the solar radiation as a fairly innocuous biometeorological agent. Yet the short wave–end of the ultraviolet can also cause skin cancer.

The geographical distribution of skin cancer in the United States shows that the southern states have a higher incidence of this disease than the northern ones. It is also established that persons with outdoor occupations are more afflicted by skin cancer on the exposed parts of the body than persons who work indoors. It is not difficult to show statistically that there appears to be a causal relation of solar radiation to skin cancer. Laboratory experiments with animals have also shown that long and repeated exposure to ultraviolet radiation will produce skin cancer. This form of cancer endangers most people with little or no skin pigmentation. Dark skin confers a good protection, apparently. Skin cancers in fair-skinned persons are evidently a result of repeated and prolonged exposure. Ultraviolet radiation is, of course, only one cause of skin cancer; there are many skin cancers not related to meteorological conditions.

Is there a way to judge the danger of sunburn without measuring the ultraviolet radiation? A rather reliable estimate for the transparency of the atmosphere is the horizontal visibility. If you can see far, radiation of most wave lengths will also penetrate well from above.

The following table indicates how the ultraviolet radiation on a cloudless day decreases in relation to the visual range. The figures are an approximation for conditions at sea level around noon. All values are expressed in per cent of those observed with a visual range of 35 kilometers or more, which is an indication of a very clean air mass.

Visual Range (in km)	35	30	25	20	15	10	5	2.5
% Ultraviolet	100	94	85	73	60	30	8	1

FOUR

Environment and Man

Our common understanding of the term "weather" is something that does not take into account the factors of the partial pressure of atmospheric oxygen and solar ultraviolet radiation. In fact, we almost never hear about these quantities in our daily contact with weather reports. They rather emphasize temperature, humidity, wind, and rainfall. These seem to govern our lives as much as those of the animals and plants around us.

In inquiring about the relations of the biological events to the weather elements one thing stands out. Weather is usually quite precisely defined and measured, but many life processes that relate to weather are often only described in qualitative terms. Other life processes can be measured with precision, but there are often links missing in connecting the two kinds of processes. Atmospheric measures are physical quantities that can be ascertained with considerable precision. Meteorologists have for many years standardized their observations. When they talk about an air temperature they mean one measured with a well-ventilated thermometer, in a shelter protecting the thermometer from direct solar radiation, at about 150 to 200 centimeters above the ground. The temperature is a direct measure of the heat content of the air.

Similarly, the wind is measured, by convention, at 10

meters above the ground. As a rule of thumb, it is well to remember that at the average height of man, 170 centimeters above the ground, wind in open terrain has 30 per cent less speed than at 10 meters. There is, however, a variety of units in which wind speed is given to the public. Radio and television weather reports generally give it in miles per hour. Aviators and mariners, for ease in navigation, prefer to have their wind speeds in knots. In biometeorological work the basic unit of meters per second (m/sec^{-1}) is in use. Factors for converting one of these units to the others are shown in the Appendix.

Temperature and wind speed are physical quantities well understood by most people. Considerable confusion, however, exists about humidity. One reason for this is that humidity can be expressed by various quantities, although usually the relative humidity is referred to. All the humidity measures are expressions for the amount of water vapor present in air in invisible form. This amount can vary widely. It is governed greatly by the air temperature; the warmer the air, the more water vapor it can carry. Cold air can carry very little water vapor. The maximum amount that can be carried at any one temperature is called saturation. It is characterized by the partial pressure exercised by the vapor. Figure 2 shows this pressure in relation to temperature. The water vapor gets into the air by evaporation from open water surfaces such as the ocean, lakes, rivers, and reservoirs. Much of it is also transpired by plants and from the soil. If air is cooled for any reason the water content may exceed the limit set by the saturation curve. When that happens it condenses, and the result is a cloud or a fog, which is a cloud that rests on the ground. If there is less water in the air than it can hold at its particular temperature, evapora-

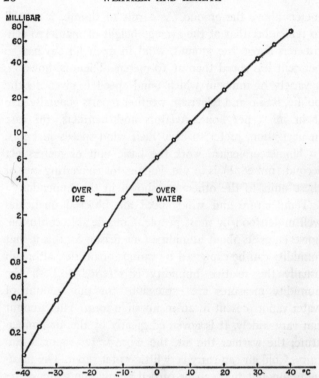

FIGURE 2. *Saturated vapor pressure diagram. The line shows the maximally possible vapor pressure, in millibars, as related to temperature; for temperatures above −5° C, saturation above a liquid water surface is assumed; for temperatures below −5° C, saturation is over ice. Ordinates are in logarithmic scale.*

tion will continue to take place from moist surfaces. The human skin bathed in sweat is one of these moist surfaces.

Although the absolute vapor pressure in air is an important physical quantity, it is more convenient to use another measure to designate the invisible moisture content of the atmosphere. This is the relative humidity. It expresses, in per cent, the ratio of the actual water vapor present in the air to the maximum it could hold at the prevailing temperature. If the air is saturated, the relative humidity is 100 per cent; if no water vapor at all is present, it is 0 per cent. In air that has a high relative humidity little water can be evaporated. In dry air, with low relative humidity, evaporation of an available water supply is rapid.

The problem in relating weather to man is that we deal always with a large group of people. In an individual person we can measure specific responses to the environment fairly well—if he is in a laboratory. But outside of the laboratory this is not very easy, even though experiments have been performed by attaching simple instruments to persons moving about in the open air. Again, the response is that of a single individual or of a small group. All these individuals, as well as the people in a large population, have their personal reactions to the environment. These reactions may be similar, but they are not precisely duplicated because of the wide differences among individuals in age, weight, body build, endocrine functions. Genetic differences play a role, too. The important point is that humans are not standardized. They are not precision instruments that react exactly the same way to environmental stimuli. If you take a series of well-calibrated thermometers and expose them to a temperature change they will, after a suitable interval of time, adapt to this change and in the end all indicate the same temperature. Not so a group of humans. Some may react by increased or decreased blood flow to the outer parts of the body; some may change their metabolism; and some, of course, may simply react by taking off or putting on more clothing. Even less than we can predict their reaction, can we tell what sensation they may experience.

Thus a wide range of responses to atmospheric changes within any group of persons has to be anticipated. There are many typical reactions, of course; a large number or majority of people in a group will experience the same reaction or sensation. But it is important to realize that we are dealing with the interactions of several very complex systems. On one side are atmospheric fluctuations; on the other, the complex responses of the human body and the even more complex responses of a large group of individuals. Therefore, many responses are established only in a statistical sense. They apply validly to a large group of persons and yet some individuals in that group may show very divergent responses. All that follows has to be read with this important reservation in mind.

All life processes are geared to one main principle. The energy intake must balance the output. Many ingenious forms have developed in nature to accomplish this aim. How does this work?

In the dim past of the earth its land masses for eons had a warm tropical climate. In it developed animals with a body temperature adapted to that environment. Food intake, which produced the internal heat, balanced the heat loss to the environment with a high degree of efficiency. Birds and mammals share this system of nearly constant body temperature, called homeothermy. Man inherited it, too. In its most basic form it can be likened to a house with a furnace. The house can be kept at a constant temperature by burning fuel when external heat losses require it. A thermostat regulates the flow of fuel to the furnace. In living beings we call this burning of fuel "metabolism." The term derives from the Greek word for change. It implies that food is changed chemically into heat and waste products. Those who are weight-conscious know only too well that we even measure our food intake in terms of a heat unit, the calorie. In passing, just recall here that a calorie is the amount of heat needed to raise the temperature of 1 gram of water (about equivalent to 1 cubic centimeter) at 15° C by 1° C. In heat-balance problems the kilocalorie is often used, which implies that we raise the temperature of a kilogram of water (about equivalent to 1 liter) by 1° C.

Long-term climatic changes and the lure of unoccupied space brought about other mechanisms of adaptation of organisms to the surroundings. In their simplest form these prevented heat loss by providing insulation. Feathers and furs became thick to match the processes that would, at an excessive rate, carry heat away from the body. In houses man has done the same thing by walls, fiber insulation, and storm windows. For the reverse effect, nature has called into action another mechanism: evaporation. This is very efficient because it takes about 580 calories to evaporate 1 gram of water at 30° C. (30° C is appropriate here as an approximate value for skin temperature.)

A look at the heat exchange between the body and the environment shows that there are three factors involved. The first is radiation. And here we deal primarily with the long-wave part. The body can receive radiation from the environment. Indoors this may be from walls or a heating element if their surface temperature is higher than the body's. Outdoors the radiation can come from the sun, the sky, and the soil. The body will also radiate toward objects in the environment that are colder than its surface. The combination yields the radiation balance, which is the aggregate of the radiative heat gained and lost. In daytime with bright sunshine the net balance is usually positive, or a heat gain. At night with clear sky and cold soil, it is usually negative, or a heat loss.

The second external factor in the heat transactions of the body is convection. This term encompasses the heat transferred to and from the body by air motion. Outdoors we usually call it wind; indoors, draft. If the air temperature is cooler than the body surface, wind or draft will carry heat away from the body. There are some conditions in which the air is warmer than the skin, for example in the hot deserts during the day or in a furnace room. Under those circumstances, the air flow will carry heat to the body.

The third factor in the heat balance is evaporation. The body always loses some heat by evaporation. The sole exception happens when the air is saturated with water vapor and fog or steam forms. We are then surrounded by water droplets and the air cannot accept any further water. But ordinarily there is some evaporation. This comes primarily from the lungs, into which unsaturated air enters during breathing. Before exhaling, the air becomes saturated and heated. The drier the air, the more water evaporates. This is a more important heat transaction than the direct warming of the air from the temperature of the surroundings to approximate body temperature. We can express this in a simple formula:

$$E_B = LMd_v \ (e_b - e_a)$$

L is the latent heat of vaporization (or as above mentioned, the 580 cal/g); M, the mass of air breathed (or about 10 liters per minute); d_v, a conversion factor for vapor density and system of units used; e_b, the saturation vapor pressure at body temperature (about 62 millibars); e_a, the prevailing vapor pressure of the air. In numerical terms, the heat loss by evaporation through breathing E_B can be well approximated by

$$E_B = 5.4 \ (62 - e_a) \text{ calories per minute}$$

At a temperature of 10° C with a relative humidity of 50 per cent the vapor pressure e_a is 6.14 millibars. Thus, the heat loss by evaporation from breathing is about 308 calories per minute. This is about four times the heat loss that results from heating the air from the environmental value to the temperature of the expired breath.

In hot weather the body perspires and further heat is lost by evaporation of sweat. The degree of cooling depends, of course, on the area exposed, the degree of air motion and its state of saturation with water vapor, as well as the total amount of sweat secretion. More details on this will be found in the later discussion of the effects of heat.

On the whole, for a normal environment the heat balance has about the following composition:

Heat loss by radiation 42%	Evaporation of perspiration 18%
Convection 26%	Breathing (a) Warming of air 2%
	(b) Evaporation of water 12%

For survival, maintenance of the core temperature of the body within narrow limits is essential. This is the basis for homeothermy. In human beings this core temperature is often equated with the rectal temperature of 37° C. The body temperature is actually a composite of the various parts. It drops at the outer surface, the skin.

In first approximation, the body temperature can be set at T_B:

$$T_B = \tfrac{1}{3}\ T_S + \tfrac{2}{3}\ T_R$$

where T_S is the skin temperature and T_R the rectal temperature. The mean skin temperature, under conditions of adequate protection, is about 33° C. With a rectal temperature of 37° C, the mean body temperature in humans is about 35.7° C. This is subject to a slight circadian, or daily rhythm of less than a degree, with a low temperature in the morning hours after midnight and a high temperature in the late afternoon, almost parallel to the daily swing of air temperatures.

We have said that life is maintained by a close equilibrium of internal and external factors. For the heat balance we can express this in a very simple formula:

$$M \pm R \pm C - E = o$$

which states that the metabolic heat (M) created chemically within the body gives thermal equilibrium if the heat gained or lost by radiation (R), that gained or lost by convection (C), and that lost by evaporation (E) add up to zero. If more heat is lost than produced, the body temperature must fall; if more is produced than gained, the temperature must rise. To prevent the former, the metabolism can be increased by muscular activity. This can be done voluntarily by work or exercise, or involuntarily by shivering. To prevent overheating, the body resorts to sweating and panting, and an increase of the skin temperature. All these factors are shown in Figure 3, which represents the basic elements of human thermostatic control.

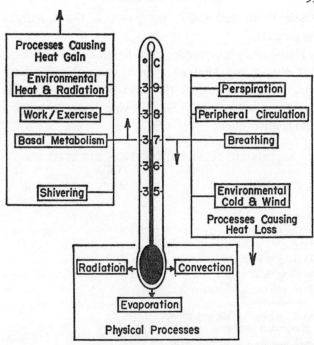

FIGURE 3. *Schematic representation of physical and physiological elements important for control of body temperature.*

For convenience the physiologists have called the heat production of 50 kilogram calories per hour per square meter of body surface one metabolic unit or 1 MET for short. This is the heat produced by a person, awake but resting. It is often designated as the basal metabolism. The body surface of man is related to his height and weight. In first approximation it can be set equivalent to a weight, so that 37.5 kilograms of body weight correspond to 1 square meter of body surface.

(More precisely the body area (A) can be calculated from the following relation: $A = w^{0.425} \, l \times 71.84$ cm^2, where w is the weight in grams; l, the length in centimeters.)

During sleep we produce even less than 1 MET heat. But almost any kind of activity raises the metabolic rate. During heavy work the metabolic heat production can rise spectacularly. The following table gives some comparative values.

METABOLIC HEAT PRODUCTION RELATED TO HUMAN ACTIVITY

Kind of Activity	No. of METS	Equivalent Kilocalories/m² hr
Sleeping	0.8	40
Awake, resting	1.0	50
Standing	1.5	75
Working at desk, driving	1.6	80
Standing, light work	2.0	100
Level walking 4 kilometers/hr, moderate work	3.0	150
Level walking 5.5 kilometers/hr, moderately hard work	4.0	200
Level walking 5.5 kilometers/hr with 20 kg pack, sustained hard work	6.0	300
Short spurts of very heavy activity (in climbing or sports)	10.0	500

Obviously, in the same atmospheric environment, a sleeping person needs considerably more protection to keep body temperature up than an athlete who vigorously competes in his sport. Thus one may find, under the same environmental conditions, a resting person covered by a blanket or comforter, while a sprinter will be happy to wear just a thin cotton shirt and shorts.

There is some evidence that individuals adjust to the climatic environment by change in their metabolic rate.

But there is still an argument among scientists about whether there is an adjustment of the shape of the native human body to the prevalent climate in a region. There is at least a suggestion that in hot climates the mass-to-surface ratio of the stature common among the people is large and thus favors heat dissipation. In contrast, in cold climates there is a tendency for a stocky build with a small mass-to-surface ratio, which restricts heat loss. Anthropologists have found from equatorial regions toward the poles a definite weight increase in the indigenous populations of the globe. 1479576

Apparently man also quite subconsciously adjusts his food intake according to the meteorological environment. An interesting test of this fact was obtained during World War II. At that time U. S. soldiers were stationed in many climatic regions of the world. They were furnished uniforms reasonably well adapted to the different heat regimes. Records were kept on their voluntary food intake. These showed clearly a much higher consumption for soldiers in the arctic than for those in the subtropical deserts. If the average daily rations consumed were grouped according to the mean outdoor temperatures, a fairly good linear relation between these two variables was established, as shown graphically in Figure 4. The difference is about 2000 kilocalories per day, with an approximate 5000 calorie consumption at −30° C, and 3000 calories at +30° C. This amounts to a little over 30 calories per degree Centigrade temperature difference. In regions with pronounced seasons people usually also adjust their food consumption throughout the year. They eat high-caloric foods in winter and much lighter meals in summer. Probably part of the increased food intake in

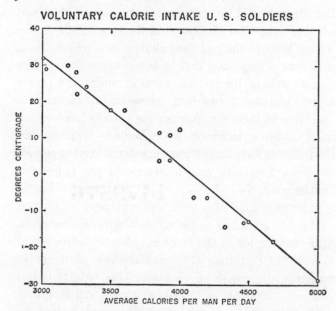

FIGURE 4. *Voluntary daily caloric intake by U. S. soldiers in World War II, stationed in various climatic zones; these are represented by the mean environmental temperature; food intake is in kilogram calories.*

a cold environment is used to produce a fat layer, which offers, in addition to a caloric reservoir, a good insulation. In fact, a subcutaneous layer of fat 1 centimeter thick is equivalent to the insulation offered by an ordinary suit.

FIVE

Coping with Cold

Man in his conquest of the globe moved gradually from the warm to the cold regions. As we have seen, through increased food intake and voluntary or involuntary muscular activity man can make some adjustments to a cool environment. He has another physiological crutch on which to lean: the reduction of the flow of blood to the periphery and consequent transport of less heat outward. This will lead to a reduction in heat loss by radiation from the skin and also make less heat available that can be carried away by convection. The extremities are most affected by this process, which operates through a narrowing of the peripheral blood vessels, called vasoconstriction. Actually we can compensate only in a limited way for external cold by this adjustment. It does help to keep the core temperature up and thus protects the most vital organs and processes. Yet cold hands and feet lead to a sensation of discomfort, and the hands may also lose a considerable degree of dexterity.

If the environmental temperature should be below the freezing point, any naked or poorly covered part of the body is in danger of frostbite. An early forerunner is chilblains, caused by blood vessel constriction and accumulation of fluids in superficial tissues, primarily in fingers, toes, cheeks, and ears. The parts become red, swollen, and painful. In the case of frostbite tissues may become frozen. Cells are destroyed and the small blood vessels become obstructed. The frozen part may become necrotic and gangrenous.

Frostbite cannot be taken lightly, for it may readily lead to the loss of limbs. In the winter campaigns during the Korean War in 1950–1951, over a quarter of the U. S. casualties were caused by frostbite. The feet are very commonly affected during cold weather as a result of the peculiar temperature distribution in the air near the ground, especially during the night hours. The temperature of the ground can be many degrees below the air temperature measured in a meteorological shelter. This low surface temperature is caused by the radiation of the soil to space. The condition with a low temperature at the surface and higher temperature in the air a few meters above it, is called a ground temperature inversion. It may become particularly pronounced when the ground is covered by snow which insulates the top of the snow cover from the steady supply of heat from deeper soil layers. On a calm, clear night the top of the snow cover can readily be 10° C colder than the air 2 meters above it. Thus a man's head may be surrounded by air at 3° C, while the heat of his feet is drained away by a snow cover at −7° C. If a soldier is immobilized by sentry duty or enemy action he may quite rapidly suffer frostbite unless adequately protected.

Protection is the key word in the race for survival in a cold environment. This race can be easily lost, even at temperatures well above freezing. A person without clothing would have to double his metabolic rate for every 8° C temperature drop. Even with the most violent shivering he can produce only about 3 MET. Actually shivering is not an efficient process. Although it increases the metabolic rate, it also brings more blood into surface layers, which in turn increases the heat loss by radiation and convection from the periphery. If we assume 30° C to be tolerable without clothing, man could theoretically compensate for a temperature drop to about 6° C, at least for a while. Some primitive tribes live in areas where night temperatures are that low, and manage to get along.

In our day it has become almost impossible to collect observations on people in an aboriginal state. Contacts with civilized cultures have provided them with clothing and other technological devices that prevent a scientific assessment of purely physiological survival techniques. But some early observations permit a qualitative assessment.

One account came from no less a personality than Charles Darwin. In 1832, during his expedition on H.M.S. *Beagle*, the ship stopped near Cape Horn, and Darwin was able to observe the living habits of the Yahgan Indians. They were naked except for a fur cape thrown around their shoulders. Darwin marveled at the fact that these nearly naked people managed to survive in temperatures ranging from 3° C at night to 7° C during the day with driving rains. He even observed a naked woman nursing a recently born baby with sleet melting on her and the child's skin. His account closes as follows: "At night, five or six human beings, naked and scarcely protected from the wind and rain of this tempestuous climate, sleep on the wet ground coiled up like animals."

Observations on Australian aborigines during the twentieth century showed that they too seemed to be able to survive the cold nights with winter temperatures from −2° to 10° C without the benefit of clothing. They use windbreaks and have small fires close to them. Apparently this bit of heat absorbed from the fire together with rapid circulation through the body parts exposed to cold makes survival possible.

A prolonged exposure without adequate clothing at temperatures near or below freezing will soon lead to a drop in core temperature. Once the rectal temperature is reduced to 32° or 31° C the person becomes unconscious. Should core temperature drop to 26° C, death will follow. This state is rapidly reached if the low temperatures are accompanied by wind, which causes a marked increase in the heat loss by convection. Each year dozens of lives are still lost by exposure in blizzards, even in the United States.

Our first line of defense against cold is clothing. It takes the place furs and feathers have in the animal world. In fact, some garments to this day come from animal hides. Downy feathers from birds were the predominant material for sleeping bags until artificial fibers came into use. The objective is, of course, to interpose an insulating layer between the body and its surroundings and thus reduce the heat loss to the environment. For convenience the people concerned with clothing have created a unit of insulation, called Clo. This is the amount of insulation needed by a person sitting quietly in a room with 21° C temperature, a draft of less than 3 meters per minute, and a relative humidity of 50 per cent. We may assume that under these conditions an individual will produce 1 MET of heat and maintain a mean skin temperature of 33° C, which will keep him in a state of comfort. One Clo corresponds to the insulation offered by an ordinary business suit. In physical terms we can define it as the amount of insulation that will let 1 kilocalorie of heat pass per square meter and hour with a temperature difference of 0.18° C between the inner and outer surface of the garment. The insulation is primarily provided by the air trapped in the weave of fibers. The best-designed garments offer an insulation of about 4 Clo per 2.5 centimeters (= 1 inch) of thickness. We can assume that our quiet person sitting in essentially calm air loses about 76 per cent of his metabolic heat through his clothing. The remainder is lost by breathing and by hand and head, radiatively and convectively.

Ingenious devices have been constructed that try to simulate the heat loss process. They are essentially metal bodies that can be heated electrically from the inside. Scientists have used spheres or cylinders or even a man-sized copper dummy to get a physical measure of the heat exchange with the environment. The tests are performed in cold chambers with varying wind speeds produced by fans. The temperature of the metallic body is kept by electric heating at body temperature under thermostatic control. The amount of current needed to keep the body warm is recorded and serves as a measure of the heat loss. These instruments have been called frigorimeters. They can be used to compare the insulating qualities of various articles of clothing. Yet the dummies only crudely approximate the intricate physiological mechanism of cooling.

For example, they convey nothing about the heat losses through breathing in cold environments. At 20° C about 8 kilocalories per hour are lost through this route. At −10° C this heat loss is doubled. Some suspicion exists that the process of heat extraction through breathing is responsible, together with other exertion, in producing heart attacks in older people exposed to cold. There is no doubt that breathing very cold air causes discomfort. For polar regions special protective masks have been designed to help in preheating the air before inspiring.

The greatest culprit, however, in cooling us down is the wind. It blows away the warm air accumulated between our skin and clothing, between layers of clothing, or clinging to the fabric. It is this convective heat loss that we feel so keenly outdoors on a blustery winter day. Even in calm air we can note it when we move against cold air through a brisk walk; or even more, when skiing fast downslope. Even low wind speeds have a very notable effect. It does not increase in a linear relation to the speed, but rather with the square root of the speed.

We can combine the ambient air temperature and wind speed to get an approximate idea of the cooling power of the air. This is often called the wind-chill index. It indicates in a rough way how many calories of heat are carried away from the body surface. Even in a wind of only 1 meter per second this is twice as much as in calm air, and three times as much at 9 meters per second. Figure 5 shows the chill index as related to temperature for some selected wind speeds, in kilocalories per square meter and hour. This is in the same units as the metabolism. A scale on the right refers to the approximate sensations experienced by a healthy young person. Any value over 300 kcal/m^{-2} hr^{-1} is felt as cool and unpleasant. When the index reaches 1400 kcal/m^{-2} hr^{-1} exposed flesh will freeze rapidly. This is obviously a highly important danger sign. It is reached with the following combinations of temperature and wind speed: $-7°$ C and 12 m/sec; $-15°$ C and 6 m/sec; $-29°$ C and 3 m/sec; $-40°$ C and 1 m/sec.

Of course, most people are not accustomed to think in terms of calories of cooling. For this reason the U. S.

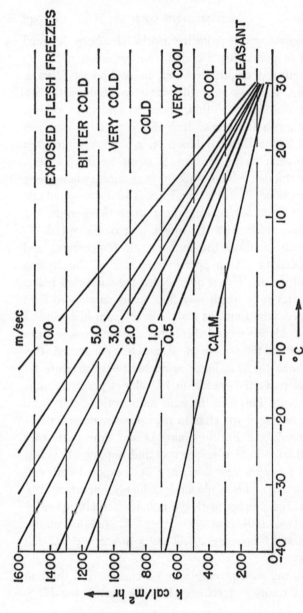

FIGURE 5. Wind chill factor in kcal/m²hr as related to temperature and wind speed (slanting lines). To read, enter diagram with temperature at bottom scale, go vertically to intersection with prevailing wind speed, read chill factor by going horizontally to scale on left. Scale on right gives approximate sensation felt.

armed forces have devised an equivalent temperature scale, representing the effect of the wind as a further subtraction from the air temperature. It reduces essentially everything to calm conditions. Lines of equivalent temperature are shown in Figure 6, for combinations of tem-

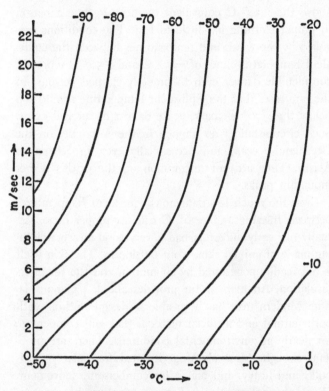

FIGURE 6. *Equivalent temperature corresponding to various combinations of temperature and wind speed, represented by curved lines. The horizontal axis gives air temperature; the vertical axis, wind speed. To obtain equivalent temperature find the intersection of temperature and wind speed in the diagram; interpolate between equivalent temperature lines.*

peratures below freezing and various wind speeds. This shows at a glance that a temperature of freezing (0° C) with a wind of 6 m/sec is equivalent to a temperature of −10° C and calm air. Or a temperature of −35° C and 12 m/sec wind would correspond to −70° C and calm. Military experience has established that in temperatures above the −30° C equivalent temperature line a person in proper clothing has little to fear. The conditions provoked by the winds and temperatures between the equivalent temperature lines of −30° C and −50° C represent considerable danger even to properly clothed persons. In the rare cases that the equivalent temperature should slip below the −70° C mark, acute danger to survival exists. Such circumstances do happen in the Antarctic, on the Greenland icecap, and occasionally even in Siberia or Alaska. They are not uncommon on the earth's higher mountain peaks.

Even the much less dangerous equivalent temperatures between freezing and −30° C call for proper dress. Actually our early ancestors had a very good answer: clothing made of animal skin, worn outside in. The skin itself is not readily penetrated by air motion and the fur offers insulation. An inch of fur provides about 4 Clo insulation. Modern man has been able to design clothing with both natural and artificial material that will compensate for nearly all environmental conditions. There are problems though. Most clothing designed for polar wear is bulky and heavy, and makes for cumbersome movement. Slight physical exertion will cause sweating, but because the clothing is nearly impermeable to wind, water evaporates only with difficulty. This can lead to ice formation within the clothing when activity ceases. Presently used face masks are still inadequate; they are apt to fog and impair vision. The glove, too, is not yet designed for manual tasks under extremely low temperatures. They are either too clumsy for handling the task or inadequate to offer protection. Perhaps space technology will soon come along with some designs that are suitable for the earth also. Man as a species has not developed a race that can naturally survive in cold regions. Only technology has enabled him to do so.

SIX

Hot Habitats

Man is well equipped to survive in a hot environment. He has, in particular, the highly effective sweating mechanism that provides for evaporative cooling in warm surroundings. Modern civilization has interposed a few problems here. We are apt to expose much less skin surface than nature has provided with sweat glands. We wear clothing even in hot environments. This clothing absorbs the sweat and is not nearly as effective in permitting evaporation as bare skin. Moreover, we have developed manners that make it impolite to sweat. So, we mop our brow to dry up sweat, defeating the body's defenses and resulting in further perspiration. We can't win on that front. Affluent people who dispose of ample electric power use air conditioning. This lowers temperatures and dehumidifies the air to bring us back to comfort without sweating.

Even though man came originally from a tropical climate, there are areas where conditions are either temporarily or permanently beyond his power of adaptation. Various combinations of the three environmental factors, temperature, humidity, and radiation, can cause heat stress. High temperatures are invariably at the basis of heat problems in nature. They can combine with high radiation intensities but low humidities to create the hot, dry desert environment. Or they may coincide with high humidities, but moderate or low radiation values to bring about the hot, humid climate of tropical jungles. Even in moderate latitudes either of these combinations may prevail seasonally, or at least on a few days. Actually the extreme conditions combining these elements are found in man-made environments more often than in nature. Boiler rooms, steam laundries, steel mills, ceramic and glass factories may have an air environment that exceeds anything found naturally on earth.

Modern ways of life also cause a considerable deviation from primitive conditions. We insist that people perform hard work under circumstances in which "primitive" man would have taken a siesta in the shade of a tree, perhaps even with a cool water puddle beside him. But industrialized man has to bulldoze roads across deserts in midday during summer, or he has to load heavy cargo on ships in vapor-burdened hot air in tropical harbors. Recent decades have also seen much military ground action in a climatic zone which severely taxes the heat adjustment mechanism of the soldiers' bodies. Unfortunately, we can't call a war off, for a few hours, because of excessive heat. So we have additional casualties among the troops in the form of heat exhaustion.

In assessing the heat load on the body, or heat stress, as it is commonly called, all the ordinary meteorological elements, especially temperature, humidity, and wind speed, have to be watched. The humidity factor can be expressed in terms other than the relative humidity or vapor pressure. For example, the wet-bulb temperature is very useful in physiological and air-conditioning problems. A thermometer that has its bulb wrapped in a thin muslin cloth is kept wet with distilled water. The wet-bulb temperature is the equilibrium temperature of a moist surface in the particular environment. If the air is saturated with water vapor the ordinary, or dry-bulb, thermometer, and the wet-bulb thermometer indicate the same temperature. No water can evaporate either from the thermometer wick or from wet skin. If the air is very dry the wet-bulb temperature will drop considerably below the dry-bulb temperature and evaporation from wet skin will be rapid. As a matter of fact, under desert conditions it is often difficult to keep the wick around the thermometer wet long enough to let it reach equilibrium with the environment. In this environment, sweat will evaporate as fast as it forms.

Another useful piece of equipment for environmental measurements in hot climates is the globe thermometer. This is a thermometer located in a hollow copper sphere, 15 centimeters in diameter. The sphere is painted flat black. It integrates various heat transactions of the environment, particularly radiation. For simulation of evaporative processes the globe has also been kept wet, to indicate a wet-bulb temperature. This is a bit tricky and in practice this quantity has usually been calculated through an approximation that relates the dry globe temperature (t_g), the dry-bulb thermometer (t_d), and the wet-bulb temperature (t_w). This yields the wet-bulb globe temperature (WBGT) as follows:

$$(\text{WBGT}) = 0.2 \ t_g + 0.1 \ t_d + 0.7 \ t_w$$

The wet-bulb globe temperature is used as an index of heat stress. Experience has shown that if it exceeds a certain level heat disorders are apt to occur. Even the tough U. S. Marine Corps has found it expedient to suspend all training when the wet-bulb globe temperature is 31° C or higher. When this index stands between 29.5° and

31° C, outdoor drill is restricted to a few hours per day and the wet-bulb thermometer indicate the same temcautions have had the beneficial effect of cutting heat casualties by 60 per cent.

Another meteorological heat index is the effective temperature (ET). The label is unfortunate because this index is not, strictly speaking, a temperature. It is supposed to indicate equivalent heat-stress values. At each given value it represents the heat sensation felt with a temperature of the same value and, simultaneously, calm air saturated with water vapor. The sensation referred to is comfort or its opposite, discomfort. The environmental conditions under which persons felt comfortable were originally established by a vote of a group of people. They were exposed in a room to varying combinations of temperature, humidity, and wind speeds. In good democratic fashion their votes were counted and for each combination of environmental factors the percentage calculated who felt comfortable. From majority votes the so-called comfort zone was established. This is shown on a diagram in Figure 7, by hatching a space. This diagram shows a lot of lines which at first may seem confusing. It is, however, a very useful graph and worth an effort to understand it. The basic grid of the diagram is temperature, on the horizontal axis, and vapor content, on the vertical axis. These two elements determine all other moisture quantities. Thus the curved lines represent the relative humidity. The dashed lines indicate the wet-bulb temperatures. The solid slanting lines are values of effective temperature. Notably, the effective temperature line of 20° cuts right through the hatched comfort zone. The majority of normally clad, healthy persons in the U. S. are expected to feel comfortable under those conditions.

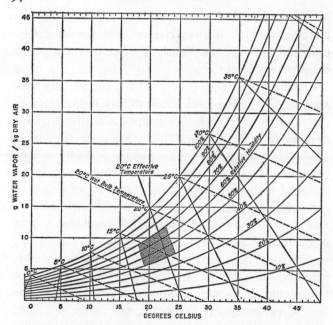

FIGURE 7. *Comfort diagram based on temperature and humidity parameters. Humidity may be expressed as prevalent vapor pressure (left scale), relative humidity (curved lines), or wet-bulb temperature. Slanting solid lines are effective temperatures. The hatched area represents the comfort zone valid for a majority of appropriately clothed people in the U. S. For interpretation of effective temperature, see text.*

Wind will shift this to somewhat higher values. At 1 m/sec the comfort zone will be about 1.5° ET, and at 3 m/sec, about 2.0° ET higher. In first approximation the effective temperature is given by a very simple formula,

$$ET = 0.4 \ (t_d + t_w) + 4.8$$

with the dry-bulb and wet-bulb temperatures measured in degrees Centigrade.

It should be clearly understood that "comfort zone" relates only to a group reaction. Individuals may have quite different notions of comfort. Very young children and old people have probably a need for higher effective temperatures than young adults. The state of nutrition and health is an important factor in the sensation of comfort. Fat and thin persons have different levels at which they are comfortable. A heavy worker with a high metabolism prefers a different climate from a sedentary person. Some of the individual differences can be compensated for by clothing. Nonetheless, the effective temperature is a fairly good index of heat stress. It shows reasonably consistent relations of the body's reactions to the heat-humidity complex.

An early reaction of the body to heat stress is a rise in skin temperature. This is obviously an attempt to radiate more heat to the environment. Skin temperature follows the effective temperature quite well. At ET 20° the skin temperature is close to 30° C. When the effective temperature rises to 30°, the skin temperature registers 33° C. For persons doing light work the rectal temperature stays normal up to an effective temperature of 28°, then it rises very rapidly. For a hard worker with a metabolic rate of 400 MET the rapid rectal temperature rise begins already at an effective temperature of 25°. Temperature and humidity combinations that exceed these values are not uncommon in some climates. Fortunately, there is some acclimatization so that people born and raised in hot, humid climates have better tolerance to life and work under those conditions than persons from moderate or high latitudes. The latter can gradually acclimate, especially if they are young and healthy. Yet while they may be able to perform usefully, they rarely consider that type of atmospheric environment comfortable.

The most important mechanism for survival in extreme heat is sweating and subsequent evaporation from the skin. Man has an estimated two million sweat glands in his skin. They can sustain a water loss of about 2 liters per hour, provided this water is replaced by drinking an equivalent amount of fluids. This corresponds to a heat loss of nearly 700 kilocalories per square meter and hour for an average-sized adult. This is certainly a very respectable compensation for adverse environmental conditions. However, such a sweat rate cannot be continuously maintained. For a twenty-four-hour interval, a rate of 0.5 liter per hour can be tolerated with adequate water replacement. Even that may not be sufficient to take care of the excessive conditions that are occasionally encountered. Under those circumstances heat sickness will ensue. In persons with impaired thermoregulation this will happen at lower stress values than in healthy persons.

There is a long list of heat disorders, some of them fairly innocent, others very serious. Among the former are prickly heat, which is a skin eruption caused by plugging of sweat ducts by keratin. Heat edema, which leads to swelling of hands, ankles, and feet, is not too serious.

Fainting associated with excessive heat is often unrelated to the water balance. It is apt to occur in people who are not acclimatized to hot environments or in those engaged in strenuous exercise. A sudden change of position from lying to standing can provoke fainting, which is brought about by a reduced blood supply to the brain, because of reduced systolic blood pressure. This is due to the enhanced flow of blood to peripheral areas. Lying down in a cool spot usually remedies the disorder.

In contrast, salt depletion of the body through excessive sweating can lead successively to fatigue, nausea, muscle cramps, and finally circulatory failure. With 0.5 gram salt deficit per kilogram of body weight only fatigue results, but if this deficit exceeds 0.75 gram per kilogram of body weight, the systolic blood pressure drops to low values and shock may result. Adequate salt replacement through food and drink in persons exposed to severe heat stress is essential.

Obviously, excessive sweating can also cause a water deficit, known as dehydration. This condition is a typical result of exposure to hot, dry weather, such as that encountered in many deserts during the daytime. For sedentary work with a metabolism at 120 MET, a temperature of 38° C, and the usual low humidity of a few per cent, 0.33 liter of sweat is lost per hour. But under the same conditions and moderately heavy work producing a metabolism of 300 MET, the sweat loss is 1 liter per hour. After a loss of about 2 per cent of body weight through sweat the person will become very thirsty. Should the loss of water run to 5 per cent, intense thirst develops and the body temperature and pulse rate also rise rapidly. Sweat water losses over 7 per cent of body weight lead to circulatory failure and may even result in death.

Body temperature and pulse rate reflect heat stress well. Both rise as the environment becomes intolerable. Acute heat disorders start when the rectal temperature exceeds 39° C. At 40.5° C there is generally complete failure of the circulatory functions and around 42° C death follows. This overheating leading to collapse or death is usually called heat stroke. Those newborn and those above middle age are most affected. In infants the cause may be as yet inadequate thermoregulation of the body, and dehydration. In older persons degeneration of the cardiovascular system is usually responsible. Heat stroke is most apt to hit persons unacclimatized to hot environments.

Heat deaths are not at all uncommon even in moderate climates. People above middle age and individuals with heart trouble are most affected. Usually their circulatory system is not geared to excessive demands of the environment. Clothing adds to the heat stress because it prevents the free play of the radiative, convective, and evaporative cooling. As little as a pair of overalls over shorts is equivalent to a 1° C rise in the wet-bulb temperature.

There are some circumstances when some clothing is beneficial. This is under high radiative conditions in the desert. It is well to remember here that exposure in full sun can add a heat load of 250 kilocalories per hour. Farmers, road workers, and soldiers are regularly exposed to this extra stress. Light clothing will reflect much of the incoming solar radiation and thus beneficially alter at least one of the elements of the heat balance.

Clothing which is too heavy for the environmental condition has caused many cases of heat prostration or even heat death. There is no clear limit of weather elements that may cause heat deaths, for it depends largely on the level of activity, duration of exposure, and state of acclimatization. We can, however, draw a line in the temperature-humidity diagram (Figure 7) that indicates the limit of danger. It is not a line represented by any of the families of lines on the diagram. It slants from the point where 25° C and 100 per cent humidity intersect, to the lower right, where 38° C and 20 per cent humidity intersect. Any condition to the right of that line calls for caution in exercise and prudence in clothing.

Strange as it may seem, most cases of heat collapse do

not occur on the hottest days of the year. They are rather noted early in the hot season, in late spring or early summer. This is evidently caused by a mixture of influences. One is wearing too much clothing; in the beginning of the warm season people have not yet shed their heavy clothing or they reason that for a single warm day it's too much of a nuisance to change suits. Another reason is inadequate acclimatization. Evidently, the body undergoes certain seasonal changes. It adapts to the warm environment gradually. In the second half of the warm season individuals seem to be better capable of withstanding extremes of heat. Dietary habits, too, change and so does fluid intake. For the heart patient early summer heat can be fatal, and he should seek refuge in air-conditioned surroundings.

Excessive summer heat is often reflected in a rise in the general mortality. A good example was the situation in the summer of 1966, in New York City. The case can be best described by use of a diagram. Figure 8 shows in the solid line the weekly death rate for the city during the year 1966. The hatched band indicates the spread in the rates that can be expected on the basis of many years of statistical records. Any point that falls outside this band constitutes an unusual happening and calls for an explanation. The dashed line indicates the weekly mean temperature for New York City. Most of the death rates are within the range of the expected but two points reach beyond it. The first in early June is not far in excess, but we note that the temperature in that week began to rise to summer values. This may have been coincidence. The rise of death cases in the early part of July is not. In the first place, deaths were far beyond the normal rate and the mean temperatures preceding and accompanying it

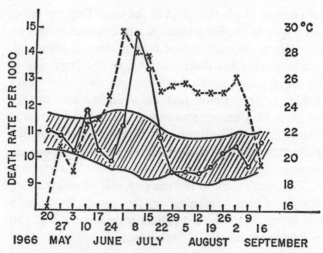

FIGURE 8. *Deaths induced by heat in New York City in the summer of 1966. The solid line represents weekly death rate (left scale), the dashed line represents weekly temperatures (right scale). The hatched band represents the range within which, from past experience, the death rate is expected to fall (95 per cent confidence limits).*

were as much as 5° C in excess of the average for the season. It was hot and sticky during that time. There is a slight lag, but part of this is caused by a difference of two days in the reporting of temperatures and deaths. It is also notable that when the temperatures dropped late in July, and stayed fairly even into September, the death rates also fell immediately to normal levels. The statistically-inclined reader will note that the upper limit of the confidence band shows a slight bulge upward from the middle of June to early July. This reflects earlier cases of occasionally higher death rates during that interval. This trend should certainly suggest precautions to physicians for their older patients.

Natives and individuals acclimated to the tropics can stand heat much better than newcomers. Casualties from heat stroke among troops operating in hot areas are always highest among recent arrivals. Particularly endangered by heat are obese persons. The rate of heat ailments among highly overweight people is three to five times larger than among the slender.

Even in outer space with very low environmental temperatures heat problems exist. Space capsules and space suits have to be conditioned for careful balancing of radiative heat gain from the sun, heat and moisture produced by the metabolism of crew members, and radiative heat losses to space. When this fails, quite unpleasant situations may arise. This was the case during Astronaut Gordon's space walk on the *Gemini II* mission. Temperatures in his space suit rose to 43° C. The humidity in his suit rose to saturation from moisture produced by sweating and breathing. This microenvironment caused physiological deterioration. His pulse rate went up to 180 beats per minute and he was unable to perform simple tasks. A look at where the point 43° C and 100 per cent relative humidity falls in Figure 7, convinces us of the danger of that situation. Only a superbly conditioned man could take it.

SEVEN

Climate, Human Evolution, and Civilization

Man's natural endowment was evidently unfit to meet the climatic challenges of all environments on earth. Yet he has managed, in contrast to all other species, to become ubiquitous. How did this come about?

The early ancestors of the human race seem to have originated about one to two million years ago. They were obviously adapted to an environment in which they could easily survive. It is not too difficult to reconstruct the climatic part of that environment. First of all, there must have been reliable supplies of fresh water. This was not only essential for early man as a drinking supply, but equally necessary for the plants that supplied his food. One can set a lower bound on the annual rainfall that would guarantee adequate, year-around water sources. Present experience would indicate that this boundary lies near 1000 millimeters of annual precipitation, with a low year-to-year variability. From the prior discussion on heat and cold, we can readily derive some limits to the temperature conditions. Man's homeothermy requires a heat balance that keeps the core temperature of the body very close to 37° C. At rest this fixes the environmental temperature at about 31° C. With some work or walking, the body produces more heat, and a level of 25° C air temperature will permit the core temperature of the body to remain constant. Temperatures that would depart very far from these values, either at the hot or cold end, could not be tolerated, except for very short times. One may set these outer limits at about 40° C and 18° C.

Hence, what one has to look for are regions of the earth where the mean temperatures are close to 25° to 30° C throughout the year, with less than about 7° or 8° C range from this mean. Obviously, there is only limited knowledge how climates were distributed on earth a million years ago. Yet there is good reason to believe that the distribution of land and water was not much different from what it is today. Nor could the general weather patterns have been much at variance with what is observed now. Geological evidence also shows that this was an era before the last major episodes of glaciation hit the earth.

Armed with this information, and with our present climatic data, let's take a stab at outlining areas on earth where human survival in the primitive stage is possible. From the above set boundaries, clearly only moist, tropical areas meet the conditions. On the map shown in Figure 9, some regions that presently fulfill the criteria are delineated. Those marked in black are particularly favorable. The hatched regions would, with only very limited protection, be suitable for human beings in a completely natural state. Actually these cover fairly sizable areas: some in Central Africa, some in tropical Latin America, and a region in peninsular and insular Southeast Asia. The last area includes particularly Malaya, Indonesia, and the lowlands of New Guinea. The paleontological record does not contradict this rather rudimentary picture. Remains of man's precursors have been recovered in African and also in Javanese deposits that are within the climatically favorable boundaries. No such finds have been made

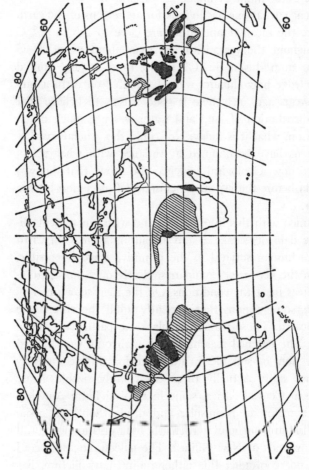

FIGURE 9. Areas with climatic conditions favorable to primitive survival. Dark areas are climatically best suited; hatched areas are suitable most of the time.

in South America. The assumption, therefore, is that man spread from one or both of the other foci.

But how has man overcome his natural limitations and spread throughout the earth? His first discovery—and one which is not original with him—is shelter. Most animal species are either using natural shelters or building them. Man must have found out quickly that a canopy of leaves protects him from rain. It also acts as a convenient shield against heat loss from radiation to a clear night sky. He would also quickly note that some bushes or fallen trees offer protection against the chilling effects of the wind. From these discoveries to a lean-to is not much of a mental jump. Undoubtedly, man would also have discovered at an early stage the value of a hollow tree as a shelter. Animals may have led him to the most desirable natural shelter of all, the rock cave. Food resources should not have been difficult to obtain in the high rainfall areas, where edible fruits and roots are abundant. Primitive weapons permitted both hunting and defense against predators at an early stage.

Yet man's struggle with the climatic environment must have started as soon as he dared to venture beyond the limits we have previously drawn. We can hardly fathom what would drive him beyond. Was it population increase or perhaps an occasional natural deterioration of climate from its optimum level? Or was it just an inherent trait to explore the unknown? Nature may have given the drift into the cold, wide world a boost. The tropical areas outlined as possible early habitats of primitive man have a high frequency of thunderstorms. Their electrical discharges must occasionally have found a combustible target and started a fire. It would take little ingenuity to note the warmth-giving properties of the embers that man might have approached. He probably also quickly learned how to maintain a fire. Only a few other inventions were needed to emancipate man from his climatic prison. The advantage of an animal skin as cover, and also incidental protection against insects, should have become quickly obvious. The use of game skins would lead both to clothing and mobile shelter. The other need was containers for the transport of water and food. Nature offers many possible prototypes of these and an attentive observer would undoubtedly spot them.

These then were the elements of a primitive technology that would permit first a seasonal and later a permanent migration to climatically less favorable regions: clothing, portable shelter, food and water carriers, and possibly a means to transport fire. The evolution of this stage of civilization may have taken tens of thousands of years, but it set man on the conquest of the earth. Reliable food supplies had to be obtained. These could not be depended upon in a primitive gathering and hunting system. Agriculture and domestication of animals had to be the next step. The raising of crops, of course, became again—and to this day remains—a highly weather-dependent activity. At the same time, in the marginal areas, man had to struggle against the onslaught of ferocious natural forces, the glaciations of the Pleistocene. These denied him substantial areas of the moderate latitudes. But it also led to a different rainfall pattern compared with the present. Areas that are now arid had then adequate moisture for plants and any crops that may have been planted. In the interglacial periods advances into previously glaciated territory were attempted but some of these forays were reversed by nature. In the last ten thousand years the climate of the present began to evolve.

Meanwhile, man's technology had made great strides. Clothing, shelter, heating, and transportation began to emerge as firmly established arts. They enabled man to settle in most parts of the globe.

However, the final conquest of the globe had to wait until this century. The last major glaciated regions of the earth—near the two poles, and the icecap of Greenland—were finally traversed. Even a few decades back this was as much adventure as it was scientific exploration. It demanded quite a few victims to the relentless onslaught of the atmospheric elements. Finally the technological advances spawned by World War II, including aviation, made routine survival and resupply possible. Shelter, clothing, and the regular transport of fuel and food have made the arctic and antarctic survivable.

A similar advance has taken place in the habitability of deserts. Where man once trekked precariously from water hole to water hole in camel caravans loaded with precious liquid, the tank truck has wrought quite a change. Even pipelines are beginning to supply outposts mostly maintained in the interminable quest for the petroleum wealth of these regions. Thirst does not threaten the workers and engineers any longer. And air conditioning is beginning to mitigate the intense solar radiation during the day.

There is now no place on earth where man cannot transport his own climate by technological devices. Yet if he were ever deprived of these, he could not survive for any length of time. His basic physiological make-up has stayed in the same state it was in an unrecorded age of primitive existence in a tropical environment. His evolution has not been, as in other species, by adaptation to the environment through genetic change. Man has rather tried to change the environment to suit his make-up. In some respects he has succeeded; in others he has created troubles for himself. Among other things, his technology has polluted the air, a theme which will shortly engage our attention.

This seems a good point to digress a little and take a look at the artificial climates man has created around himself. In this technological age there is nothing to prevent us from having atmospheric conditions around us that suit our purposes best. Even the most primitive type of housing eliminates quite a few unpleasant weather conditions. A roof will stop all types of precipitation. Add walls and you have complete protection from wind. Modern houses are constructed to withstand all the weather vagaries except perhaps the most catastrophic. Fortunately, the probabilities of being struck by a tornado or hurricane, even in areas where these weather extremes occur, are very small. Better building codes and better materials continuously decrease the residual danger.

In other respects, too, houses decrease the impact of weather elements. Ultraviolet radiation is completely eliminated. Windows will not let any of the short wave lengths pass through. Light intensity is also radically diminished. The subdued light indoors is obviously more agreeable to the human eye than intense sunshine outdoors. The total radiation can, in warm climates, cause a considerable heat load on the buildings. All building materials absorb it. Roofs are particularly apt to act as heat traps. Most roofing materials are poor reflectors and in bright sunshine become 10° to 20° C warmer than the air. This heat is transmitted through the attics into the houses. Glass surfaces, so dear to modern architects, also reflect very little radiation. In fact, they trap much of it, because they are impermeable to all infrared wave lengths that radiate back from the inside. Of course, in cold climates every bit of sunshine and radiation is appreciated as additional heating. In hot, sunny climates, experience has taught people to reduce window surfaces, or even completely omit them on the outside. Arabic construction favors such designs. Sunny Spain is the origin of the house with few windows on the outside and on the inside the tree-shaded patio, into which the windows open.

Other construction methods help by insulation to even the wide temperature swings between day and night, summer and winter. Insulation simply dampens all the oscillations and makes it easier to maintain a different and more even climate indoors than outdoors. In the Middle Ages people accomplished this by thick walls, which were not only a defense against militant intruders, but added to the thermal inertia of the houses. Today, we accomplish the latter defense by glass wool and storm windows.

This helps our heating and air-conditioning plants by reducing the load imposed by the wide fluctuations of weather. We can illustrate them best by the temperature swings. These can be regular, as between day and night or summer and winter; such variations are called periodic because the extreme high and low values are either about a day or a year apart. Other wide swings are quite irregular in timing; they are caused by air flow from different quadrants and the passages of weather fronts, which separate warm and cold air. These changes of temperature are called aperiodic because they can happen at any time.

The daily periodic swings are largest on clear, sunny days, especially on the deserts. A difference of 30° C between the lowest value just before sunrise and the highest value in the early afternoon is not unusual. In contrast, on oceanic islands the change throughout the day may only be 2° C. On overcast days these daily swings become also very small in other localities. Between midsummer and midwinter we can measure the periodic swings by taking the difference between the mean temperatures of the warmest and the coldest months of the year. In the middle latitudes on the Northern Hemisphere these are usually January and July. Two contrasting examples in the United States are values for San Diego, California, and Bismarck, North Dakota. In San Diego this range, between January with a mean temperature of 13° C and July's 20° C, is only 7° C. The corresponding values for Bismarck are January, −23° C; July, 22° C; hence a range of 45° C. The former city is located in an even maritime climate; the Great Plains city, in a continental climate that has much wider limits.

The aperiodic temperature changes also show broader sweeps in mid-continent than at the west coast of the U. S. These changes, brought about by storm systems, are readily noted in the difference of temperature from one day to the next. Even in the winter season, these aperiodic differences are small in San Diego where they average only about 1.5° C. In Bismarck they are about 5° C between successive January days. On the eastern seaboard, which is partly under continental and partly maritime influences, they run between 3° and 4° C. Within extreme weather changes the rises and falls of temperature in San Diego stay less than 12° C, whereas in mid-continent changes of 22° C within twenty-four hours are not uncommon. These large changes impose considerable loads on the body's capabilities to adjust. More about these meteorological stresses follows later. Here

these changes illustrate the forces our houses and mechanical devices have to counteract.

Experience has set the standards that should be met
indoors, varying somewhat from country to country and
season to season. They depend on clothing habits and
purpose of the structure. In residences in the United
States the so-called ideal is 21° C, a relative humidity of
50 per cent, and at least two complete air changes per
hour without undue draft. The farther the outdoor conditions are from those desired, the greater the need for
mechanical devices that can compensate for the deficiencies. Man's varied ingenuity has met the climatic
problems in varied ways, be it by caves, Eskimo igloos,
or early dwellings with old-fashioned potbellied stoves.
It is sufficient to say here that the technologists of the
twentieth century have designed heating and cooling
plants to meet all needs, albeit at considerable cost in
energy. For example, to be comfortable indoors at the
standard air temperature of 21° C and a wall temperature
not below 18° C, we must start heating when daily mean
air temperatures outdoors drop below 18° C. For a home
of average construction there is a need of about 3 kilowatt
hours per day for each degree centigrade the temperature
drops below 18° C, for every 100 square meters of floor
area. This energy is obtained from coal, oil, and elec-
tricity, but some efforts have also been made to harness
solar energy for heating purposes. At 25° C, especially
with humidities at or above 60 per cent, the need for
air conditioning arises, with energy requirements of
similar magnitude as for heating.

Problems still exist in rooms, except for those equipped
with the most modern heating and cooling devices. One
of the problems is vertical temperature stratification.

Floors are usually cool and ceilings are warm because warm air rises, or, as we have already seen, in summer, heat seeps through from the roofs. For optimal comfort the temperature should be distributed the other way around, warm underfoot and cooler overhead. In wintertime heat is often drained away through our feet unless we take pains to minimize this by insulating materials, such as carpets or cork tiles. The problem is particularly encountered in schoolrooms, where students often have to sit quietly for considerable periods of time. Yet the ancient Romans already had house designs with underfloor heating.

The comfort zones in work rooms and factories are, of course, quite different from homes. Many of them should be considerably cooler than residences because of the much higher metabolic rates of workers. The desirable conditions are often not met. Environmental problems are particularly encountered near ceramic and blast furnaces, foundries, and in steam laundries. Workers in some of these trades, aside from needing large amounts of fluids, have to be rotated after short exposures to cool rooms for rest periods.

There are also special needs to control the atmospheric environment for the very young and very old. For sick persons, particular protection against environmental stress is often indicated.

It is worth a moment's thought about the air environment in bedrooms. A good bit of our lifetime is spent there. When resting or sleeping, only very light clothing is worn and the metabolism is low. This suggests the need for very even and comfortable conditions. In cool weather, with proper heating, this is easily accomplished. A temperature of 20° to 22° C is widely acceptable as standard. Younger persons may prefer it a degree or two colder; old people, generally, like it about 2° C warmer. In summer, at temperatures about 28° C, or vapor pressures above 21 millibars, bedrooms become very sticky. Air conditioning is the remedy for some, but a night spent outdoors will quickly educate us to nature's way of cooling. It may be radiation to the sky or a slight breeze, or a combination of the two. City dwellers are, unfortunately, often deprived of opportunities to use the outdoors that way.

For particular comfort during rest or illness man resorts to the bed. Here, through blankets, quilts, and feather beds, he separates from the room a small space where he is surrounded by a particularly protective climate. Under the covers the temperature fairly quickly adapts to the temperature of the skin. If the bed is cold before it is occupied, the skin temperature and bed temperature will be fairly low, about 3° C. If the old custom of preheating the bed is followed, the temperature of both skin and bed will be around 33° C. Electric sheets or blankets, controlled by a thermostat, are the modern answer for comfortable warmth in bed. Even with room temperatures as low as 10° C, they will maintain under the blanket close to comfortable skin temperatures. No comparable cooling devices exist, although water-cooled plastic mattresses exist. These float on a water bath and are designed for cooling sick persons with high fevers.

EIGHT

City Climate

Beds, rooms, and dwellings are a very natural extension of man's primitive adaptation to climate. They are not very far from a spread of dried grass and a cave. Even a small group of abodes is in harmony with nature. But if the tribal village grows into a town and a town into a city, a major change takes place. Accumulations of houses, business and apartment buildings, streets, and parking lots have caused a major change of the landscape. Add traffic and industrial activity to this, and forces are set into motion that alter even the atmosphere radically. Clearly man did not plan it that way. It just happened. With the world population on the increase, still more natural land than in the past will be transformed into a new barren expanse. As fewer and fewer people are needed to man our mechanized farms the cities will continue to grow. All this has brought about social, aesthetic, and physical changes. Here only some of the physical changes shall concern us, yet it shall be apparent that they have important relations to the aesthetic and social factors rampant in urban life.

Just what has man wrought? First of all, he has replaced the soil and plant cover by stone and asphalt. He has arranged to have natural rainfall drain away as rapidly as possible. Enormous amounts of heat are produced by the massed bodies of people, by heating, by industrial combustion processes, and by vehicles. Industrial activities and propulsion engines are adding water vapor, various gases, and solid substances in huge quantities to the air. The modern city has been likened to a desert with volcanic action, continuously belching forth noxious effluents and dust.

What are the effects? Let's first look at the changes that have been brought about in the weather elements. Some can be noted without instruments. Solar radiation has been reduced. The layer of haze above the city absorbs it. The total radiation, as received on a horizontal surface, is 15 to 20 per cent less than in the rural environs of a city. Ultraviolet can hardly penetrate in the cooler season and is markedly reduced during the warmer season. Occasionally, however, cities are blanketed by such a heavy pall of smoke that in midday street lights are needed.

Simultaneously, the metabolism of all the people and the innumerable combustion processes going on in a city add heat to the air. In some of the metropolises of the moderate latitudes, this amounts to about a third of the heat received during the year from the sun. In cities with narrow streets or deep canyons formed by tall buildings, the outgoing radiation is also quite different from that of a flat grass surface. The walls of the buildings radiate toward each other, instead of toward the sky. Their high heat conductivity brings warmth from the interior outside. Their high heat capacity stores heat received from the sun during the day and gives it off only slowly at night. All this keeps the city warmer than the countryside. The difference is relatively small in midday. It is largest shortly after sunset, but persists throughout the night. This temperature difference causes the urban "heat island," as that effect has been labeled. A typical example is shown in Figure 10. The temperature difference is closely related to the population density, and the packing of houses, with the densest settlements showing the most heat surplus. There is one beneficial aspect to the higher city temperatures. In regions with snowfall, snow covers will form later and melt earlier in the cities than in the open country. This is a small consolation prize, in an

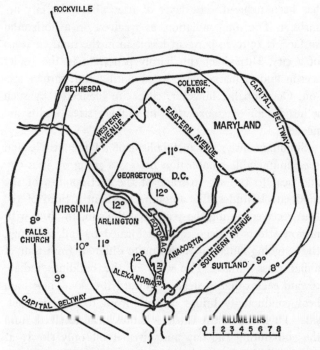

FIGURE 10. *Typical nocturnal temperature distribution in the Washington, D.C., metropolitan area. Lines show approximate isotherms. Highest temperatures are shown in the densest built-up areas of the District of Columbia, Arlington, and Alexandria. Park areas and suburban sections are coolest.*

otherwise losing game for a beneficial biometeorological environment for human beings.

The wind, too, is considerably affected by the buildings of a city. They act as obstacles, reducing the wind flow and robbing the city of ventilation it badly needs. The records show that wind speeds in the city are reduced 20 to 30 per cent compared with the free flow noted at the nearby airports. At the same time rising air currents produced by the surplus heat are found over cities. They may even lead to increased showers in summer. This tendency is reinforced by the millions of tons of water released into the air from steam-power plants and innumerable combustion processes. In spite of all this the relative humidity is generally lower in urban areas than in the suburbs and the country. The soil and the plants which act as sponges for rain are missing. Hence the slow but continuous evaporation which brings water vapor back into the air layers near the surface is absent. Storm sewers rapidly remove the moisture.

However, all these effects on temperature, wind, and humidity dwindle into insignificance compared to the dust and the contaminants a city brings forth. The following table gives some startling figures on what is annually belched into the air in the United States, mostly in urban areas.

None of these freely flowing "gifts" of our industrialized civilization are pleasant and some of them are definitely dangerous to health. No housewife will cheer if a dustfall of 30 tons per square kilometer per month soils her laundry, and blows into her house and blights its paint. Carbon monoxide in the concentrations occasionally found near heavily traveled thoroughfares of the city may cause headaches. Sulfur dioxide is probably the worst

IMPURITIES ADDED TO THE AIR ANNUALLY IN THE UNITED STATES IN MILLIONS OF TONS

	SOURCES			
Substance	Refuse Disposal	Heating	Power Plants	Motor Vehicles
Solid dust	1	1	3	1
Carbon monoxide	1	2	1	66
Nitrogen oxides	1	1	3	6
Sulfur oxides	1	3	12	1
Hydrocarbons	1	1	1	12

of these unwanted admixtures. Even in minute doses it constricts the breathing passages. It has been blamed as the culprit in some of the major air pollution disasters that have occurred, killing hundreds of people, and very subtly contributing to the death of others. Very likely sulfur dioxide was only one of many pollutants acting in concert.

Usually the health hazards of air pollution are hidden in the maze of the many illnesses affecting man. Occasionally, though, these hazards reach such disastrous levels that they cannot be swept under a rug whose fabric is ignorance, indifference, and inertia. One such case claimed thousands of human lives. It occurred in London, England, in December 1952. In the calm, cold English climate of the winter season, hundreds of thousands of furnaces and fireplaces belched smoke from sulfurous fuels into the air. At the same time, a stagnant air mass, in a big high-pressure system, settled over England. Wind motion was small. An inversion of temperature caused the cold air to hug the ground and prevented mixing with the higher air layers. A damp fog formed into which the sulfur dioxide fumes were discharged. They partly reacted with the fog droplets to form sulfuric acid, which is

even more of an irritant than sulfur dioxide. The stagnant
air stayed for several days. The sulfur dioxide concentration rose to over five times the initial value. And with the
accumulation of SO_2 rose the deaths in the population.
Both the change in the strength of the noxious gas and
the number of daily deaths is shown in Figure 11. For four
dismal days, the sulfur dioxide accumulated and deaths
soared. Even after the pollution levels returned to normal

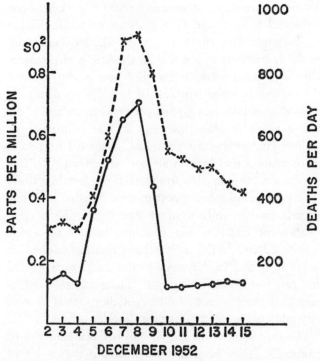

FIGURE 11. *Deaths in London, England, during atmospheric stagnation period in December 1952. The solid line (left scale) shows sulfur dioxide concentrations; the dashed line (right scale) shows the number of daily deaths.*

the death figures stayed above the usually expected level. The big city had suffered a fumigation usually reserved for pests, and the weakest members of society had succumbed. The only consolation is the fact that a Clean Air Act was promptly passed by Parliament to help in the prevention of similar disasters. Yet in many of the world's growing cities much remains to be done to stave off similar events.

Because of its dangerous influence on the human respiratory and cardiac systems much research work has been performed to establish what amounts of sulfur dioxide are harmful. The effect, of course, depends not only on the concentration, but also on the length of exposure. Thus very small amounts breathed over a long interval of time can be quite harmful. In fact if a population is exposed to as little as 1.5 parts per hundred million of air over a year's duration, increases in the occurrence of cardiorespiratory diseases are noted, compared to a similar population that lives in air not contaminated by this gas. On the other hand, a strong whiff one hundred times more powerful of this pungent mixture, inhaled for a few seconds, may do no harm at all. The threshold of a combination of duration and concentration, above which definite adverse health effects have been established, is shown in Figure 12. Any combination that lies above the line can be considered harmful. There are also indications that the presence of other pollutants, such as soot, may add to the effects of sulfur dioxide.

The oxides of nitrogen and the hydrocarbons are of similar ilk. They have the bad attribute that a bit of ultraviolet solar radiation will change their chemical composition through photochemical reactions. Some of them will transform into ozone. This triatomic oxygen is a

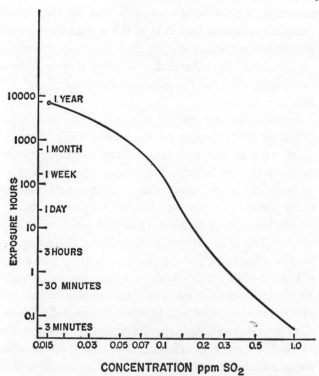

FIGURE 12. *Threshold concentrations of sulfur dioxide considered to be harmful to human beings for various exposure times (left scale, logarithmic time units).*

notable irritant to mucous membranes. Other ill effects, such as nausea, have been ascribed to even small concentrations. The indescribable chemical witches' brew resulting from the photochemical reaction of auto exhaust gases leads to the well-known smog. This term graphically describes a condition halfway between smoke and fog. It gained particular notoriety in the Los Angeles area, where free air circulation is impeded by orographic conditions (mountainous terrain), and by the vertical temperature distribution. Add to this a good dose of solar ultraviolet and exhaust gases, and you have everything needed for a photochemical caldron. The most obvious immediate effect of the resultant chemicals is eye irritation.

Yet this does not tell the full story. There is a slower action of air pollutants on the human body. Some of them, such as lead and arsenic compounds, accumulate in the body and imperceptibly poison the system. Even more insidious are some cancer-causing agents that are quite common in city air. Are they responsible for the higher lung-cancer rates in cities compared with rural locales? This is not known with certainty, but a finger of suspicion has to be pointed at them. There is also a high sickness rate in the city from bronchitis and emphysema. The latter causes a marked reduction in lung volume and makes breathing difficult. It is more common in cities than in rural areas. Air pollution, next to smoking, seems to be the culprit in affecting the inhabitant of cities more than his farming counterpart. The statistics leave little room for doubt that other respiratory diseases are aggravated in the city environment at times when the sulfur dioxide content of the air is high.

In some sensitive persons air pollutants provoke asthmatic attacks. They disappear when these individuals move from the area. The specific chemical that triggers the attack is usually unknown. With hundreds of compounds present that are foreign to clean air it is difficult, if not impossible, to single out what causes these essentially allergic reactions in people. For this reason, some cities have gained the doubtful distinction of having their name associated to these ills: Yokohama asthma or New Orleans asthma.

Another problem closely related to meteorological conditions also plagues our cities. One word describes it: noise. The origin of the noise is, of course, unrelated to weather. It may be traffic, the ubiquitous jackhammer, or low-flying aircraft. In all these cases the noise is carried along and eventually dissipated in the atmosphere. Two closely linked factors essentially govern noise propagation. One is the wind speed and turbulence, the other is the vertical temperature stratification. We can characterize two distinctly different conditions. In the first we have a rapid temperature decrease with height above the ground; the wind is brisk and has many little eddies in it. Under this set of circumstances, sound is rapidly dissipated and doesn't even carry very far downwind. But let the temperature increase with height above the ground, and imagine a very slow steady wind, and you have the second set of conditions. Under these circumstances, the sound is ducted and stays close to the ground. You may even hear a railroad whistle for miles. When this atmospheric pattern exists near the busy jet airport of one of our cities most people in the neighborhood will voice objections. High noise levels are definitely irritants and have to be counted among the atmospheric pollutants of modern city environments.

City structure has a great deal to do with city climate. Much can be done to bring about improvements. One thing to avoid is an extended mass of houses, crowded together and projecting a uniform level of roofs. This essentially creates a new surface for the atmosphere to move over. The wind does not ventilate well and radiative exchanges below are slowed. An open lattice of houses and buildings with wide streets, with roof heights in the neighborhoods at many levels, large parks and green spaces would do much to improve the atmospheric hygiene of cities. Obviously, both technological and legal steps are necessary to contain and reduce air pollution.

The following table summarizes the changes in the atmospheric environment brought about by urbanization. These apply to cities in moderate and higher latitudes. Little quantitative knowledge is available about cities in the tropics.

CLIMATIC CHANGES PRODUCED BY CITIES

Atmospheric Element	Change Compared with Rural Areas	
Radiation, total on horizontal surface	reduced	15–20%
Ultraviolet	"	10–30%
Sunshine duration	"	5–15%
Temperature (nights)	increased	1–3° C
Heating needs	reduced	10%
Relative humidities	"	2–10%
Rainfall	increased	5–10%
Number of rainy days with small amounts	"	10%
Snowfall	reduced	(amount depending on latitude)
Cloudiness	increased	5–10%
Fog and low visibility	"	50–100%
Wind speeds	reduced	10–30%
Contaminants (solids)	increased	1000%
(gaseous)	"	500–2500%

NINE

Weather, Performance, and Behavior

The influences of the atmospheric environment on man presented so far all deal with direct physiological reactions. A weather element, such as temperature, changes, and the body tries to adapt to the change. These effects can be duplicated in experimental chambers in the laboratory but there are apparently more subtle and indirect influences of weather. These are not easily explainable, and laboratory duplication has been possible only in a very limited fashion.

In these cases there is essentially a wholesale effect of the weather on the organism. Many atmospheric elements change at the same time and the response of the body is more subtle than in cases of simple heating, cooling, or deprivation of oxygen. In studying these effects scientists have used weather patterns as a correlate to the biological events. In the latitudes between 20 and 60 degrees the migratory pressure systems control the weather patterns. These systems, usually characterized by their barometric pressure, chase each other in irregular, unending sequence. Systems with low pressure, the cyclones, are commonly called storms. They range in size from a tornado about 30 meters in diameter, to a hurricane of 100 to 200 kilometers diameter, to winter storms that may measure up to 1600 kilometers across. These cyclones have one thing in common. The winds circulate counterclockwise around them (clockwise in the Southern Hemisphere). Most of the cyclones also move from west to east. Therefore on their front sides they have winds bringing warm tropical air from lower latitudes while to the rear they have winds that bring cold polar air. Cold and warm air masses do not mix where they meet but they interact. In fact, their interfaces are the most active weather zones. They cause clouds and precipitation. The change from one air mass to another is usually rapid and takes place in a short horizontal distance. The interfaces between these different bodies of air are called "fronts." In frontal zones much of the violent weather occurs, including many thunderstorms.

Between storms an anticyclonic pattern prevails and the weather often quiets down. The pressure is high and winds are slight. Around these high-pressure areas the winds on the Northern Hemisphere circulate clockwise (in the Southern Hemisphere, counterclockwise). Skies are clear, and bright sunshine is the rule.

Biometeorologists have made much use of a simplified weather scheme. It breaks the common sequence of weather events in moderate latitudes into six phases. This model is shown pictorially in Figure 13. It depicts a typical sequence of weather from right to left in the familiar symbols of the weather map. At the bottom this diagram shows the usual changes in meteorological elements that accompany the passage of the weather system. For convenience each weather phase shown here will comprise one calendar day. Actually, these phases are of irregular length. The totality of weather events and changes in all elements simultaneously is assumed to have the biological effect. This model will serve in all following discussions.

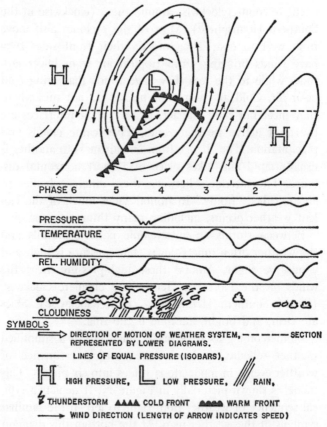

FIGURE 13. *Schematic representation of weather phases that are related to biometeorological events. Each phase in the diagram spans twenty-four hours. In reality the phases may be shorter or longer.*

Let us first turn to the far-reaching influences of weather on performance and mental attitude. The simple, adverse effect of hypoxia has already been presented, but is, of course, an uncommon occurrence in daily life.

It is also, of course, readily clear that a person in a comfortable environment will perform his tasks better than someone who is acutely uncomfortable. Many tests have shown this. Let me cite some. Telegraphers make considerably more mistakes when surrounding temperatures go over the 32° C mark. Typists, similarly, transcribe their material more slowly and make more errors when they are exposed to an uncomfortable atmospheric environment. Factory work performance shows lower production rates and poorer quality when heat or cold act as irritants.

A large number of comparative tests have also been performed on school children. Carefully matched groups have been alternately placed for several weeks in rooms that were heated and humidified during cold weather or air-conditioned in warm weather. In either case, indoor conditions were precisely placed in the zone of optimal comfort. The control group was in ordinary schoolrooms, heated when necessary, but not air-conditioned. Progress after several weeks and standard test scores were compared. The children in the rooms with carefully controlled atmospheric conditions not only learned better, but also made fewer errors in their tests. When the control group was moved to the rooms with climatic conditioning and the other group to the ordinary rooms, the scores reversed. Again the children in the controlled environment did better in their studies. The differences were greatest during times when air conditioning was needed. Evidently discomfort at the warm end of the thermal scale interfered with the learning process. Perhaps it was only the yearning for the old swimming hole that caused the differences. But the reasons are likely to be more involved.

Obviously our mental attitude and our attention span change with the changes in the air environment. There is reason to believe that this is not entirely an effect of the constellation of temperature, humidity, and wind. More subtle forces may well be at work at the same time. Some scientists have given such attributes to electromagnetic waves, originating from thunderstorms. Others have invoked the ions in the atmosphere. These are small charged particles in the air, which owe their existence to the cosmic rays that hit the earth from the universe, or to decaying radioactive substances present in profusion in soil and rocks. These ions, especially when they have just formed, are small and highly mobile and are assumed to be biologically most active. Later in their life cycle they become attached to larger aerosol particles and are more inert. How and why the electromagnetic waves and ions act is still a mystery. Answers are being sought in tedious animal experiments, which up to now simply confirm that there is an effect. So for the time being, we have to be content with the skimpy knowledge of an influence without knowing the precise mechanism.

One hypothesis surmises that there is an interaction between fluctuating electric fields and human brain waves. The latter can be monitored by the electroencephalogram, a technique that records electric fluctuations in the brain through electrodes suitably placed on the skull. They show when the person is in normal waking state and when he becomes drowsy or falls asleep. Tests on human beings are usually brief, a few hours at most. In order to test weather influences one would have to keep many people under surveillance for a long time. This is manifestly an experiment that is unlikely to be performed. So, all we have are a few sporadic samples. These do suggest that electric field changes and fluctuations may in some persons lead to sleepiness. This is probably also the explanation of the correlations that have been found between barometric pressure and wakefulness. The barometric pressure likely is only an index of various states of the weather. Frontal conditions, which are associated with rapid pressure changes, are often also linked with electrical field changes, including thunderstorms.

If we assume that these electric effects on the brain are real, it is not difficult to understand other observations to the effect that human reaction times show a link to weather. At a traffic exhibition in Germany, twenty thousand visitors were tested for this trait in an interval of ten weeks. Reaction time was checked in a simple setup, in which a person operates a push button when a traffic light in front of him changes color. The experimenters noted considerable differences from day to day, quite aside from the notable variations between individuals. Yet it became quickly obvious that the whole group spectrum on certain days showed greater alertness compared with other days when everybody seemed to be more sluggish. When the observations were arranged according to weather phases (Figure 13), things began to make more sense. On days when weather phases 1 and 2 prevailed, the people tested show fast reaction times. During phases 3, 4, and 5, slow reaction times prevailed. In weather phase 4 a fairly large number of persons were particularly slow. In phase 6 the values of reaction time were close to the average for all tests. Weather phase 4 is associated with poor weather and often has electrical disturbances. All the data were carefully analyzed by modern statistical tests so that little doubt about their validity exists.

Is there an answer here to some unexplained automobile accidents? Motorists usually slow down for obvious weather hazards, such as snow, ice, and wet pavement, but no one as yet warns them about the imperceptible weather elements that may dull their senses.

Yet these subtle weather effects seem to extend into industrial accidents as well. These show a pattern very much like the reaction time. In an industrial plant of six thousand workers, during twenty million wage-hours several thousand accidents occurred. Many of them were minor in nature, but nonetheless work time was lost. Here again days with weather phases 3 to 5 had an above-average number of accidents, but phases 6, 1, and 2 showed fewer accidents than the average of all days. Reaction time changes undoubtedly contribute to this distribution of accidents. Other research indicates that thermal discomfort is reflected in industrial accidents. These studies show that when air temperatures exceed 24° C accident rates rise. Similarly, when the air temperatures go below 12° C accidents occur more frequently than in the intermediate temperature bracket.

Not all the information on behavioral problems is as yet coded according to weather phases. Many of the past studies simply record pressure, temperature, and humidity. These elements have been related to classroom behavior of grade-school students. Restlessness and infractions of discipline were noted by the teachers again with high temperatures and humidities. By inference, the meteorologist would often associate such conditions with weather phase 4.

Similarly, there are some studies that relate suicides to rapid pressure changes. Here again the pressure is merely an index for the whole system of weather patterns. In the usual course of weather events rapid pressure changes are associated with frontal activity. When that type of weather prevails, all individual elements fluctuate rapidly. Why weather phase 4 in all likelihood should lead to mental depression, which may end in self-destruction, is not known. We merely observe the end results.

Some scientists have also linked the occurrence of riots to weather. The term "long hot summer" has acquired this connotation in folklore. Again, that irritating combination of high temperature and high humidity seems to provoke the abnormal conduct. There may be nothing more behind it than the fact that old, inadequate housing causes such an oppressive atmospheric environment to drive people into the streets. Cold and rainy weather, on the other hand, keeps people indoors. However, we cannot exclude the possibility that the atmospheric environment may do more than cause physical discomfort. There appear to be equally strong forces at work that affect our mental attitudes. In the countries of Southeast Asia, the period just before the summer monsoon is known to cause persons to run amok. They seem to lose control in their senseless frenzy to kill. With the cooling monsoon rains, a feeling of tranquillity returns.

Occasionally, nature performs an experiment for us. This happened at Massachusetts State College in September 1938, during the passage of the famous New England hurricane. The entering class of freshmen took three standard tests, one on the day before the hurricane, the second during passage of the hurricane, and the last on the day following the hurricane transit. In the first test, results were slightly above average; in the last test they were a bit below expectations. That is nothing startling, but the scores for the test taken during passage of the hurricane when the pressure first dropped rapidly, and then rose again abruptly, showed a score 20 per cent above the usual performance for students at that institution.

There has been no repetition of such a coincidence. Hence we should not rush to proclaim that hurricanes raise intelligence performances. Yet we should not discount the power of an abnormal weather combination on mental performance either. Sustained critical experiments remain to be done; the questions they should answer are, however, well established.

TEN

Weather Suffering

The notion that weather influences disease is very old and embedded in the folklore of many civilizations. But nowhere did it reach as much rationalization as in the classical antiquity of Greece. A brief mental journey to the little Greek island of Kos, just off the coast of Asia Minor, will show us where it all started. Let the time be about 400 B.C. or nearly twenty-four hundred years ago. That was the heyday of the hero of this story, Hippocrates, who lived approximately from 460 to 377 B.C. He is, of course, the most famous physician of the ages. But first it is appropriate to say a word about Kos. Its claim to fame was the temple dedicated to Asclepius (alleged to have lived 1321–1243 B.C.). Asclepius, deified by Greek mythology, was an early practitioner of the healing arts. His habitat in Kos became what we would now call a medical center. Sick people would flock there to receive herb medicines, baths, and other treatment. Descendants and pupils of Asclepius learned their master's secrets and kept the center going. They built up a strong tradition of medical art.

Little is known about the patient's obligations in the ancient Medicare system. Yet for those who recovered their health it was a duty to present a testimonial. The healed visitor to Asclepius' temple had to record the type and symptoms of his disease and the therapy which brought relief. These were written on small tablets and, in honor of the deity, hung on the columns of the temple. It is likely that the physician-priests have kept some records on those who did not make the grade.

In this world of empirical medicine Hippocrates took up the physician's art as a descendant of Asclepius in the nineteenth generation. Temples to his ancestor had been erected elsewhere and as "Asclepeions" were the clinics of the era. Not many were rivals of Kos in reputation. That was the Mayo Clinic of the age. Evidently, Hippocrates' predecessors had studied the little tablets with case histories and drawn some conclusions on causes of diseases, their course, and the effect of therapy. Hippocrates seems to have absorbed this medical experience of centuries at an early age and become a skilled master of the healing arts. He also traveled and visited other Asclepeions. This gave him a chance to obtain more medical information about the cumulative experience of other clinics.

The great step taken by Hippocrates, or at least ascribed to him, was the compilation of reviews of the knowledge about various phases of medicine. These books were the texts of all the accumulated experience and for centuries were considered the acme of medical science. They have been translated and retranslated and commented upon to this day. Also ascribed to Hippocrates is an encompassing syllabus of all the knowledge he had reduced to writing. These so-called *Aphorisms* and one of his books, *Airs, Waters, and Places,* are pertinent to our theme.

In them are found for the first time organized comments on the effects of instantaneous weather and seasonal weather conditions on disease. In fact, Hippocrates states boldly that one should start the study of medicine by considering the relation of seasons to disease, and next investigate the medical consequences of various winds. He succinctly states that hot and cold winds have quite different bearings on the outbreak and course of diseases. Then he lists diseases and symptoms prevalent in various seasons.

Here are some of his views: Dry years are, on the whole, healthier than wet years and show lower mortality. Among the diseases believed to be promoted in rainy seasons are protracted fevers, gangrene, epilepsy, apoplexy, and quinsies. In dry weather consumption, arthritic ailments, dysentery, and eye inflammations are prevalent. Hippocrates also comments on individual weather situations, which he astutely relates to the wind prevailing at the time. When he mentions a north wind there was likely to be cool weather, while a south wind indicates warm weather. With north winds he connects coughs, throat ailments, constipation, and pains. But he also notes that the northerly currents are bracing, tone the body, and improve hearing. The south winds, in contrast, dull hearing, dim vision, and lead to weakness and inactivity. They are also alleged to loosen the bowels, and cause headaches and vertigo. Interestingly enough, he also wrote that some people react just oppositely to others in weather changes. He explains this by the different body build and temperament of various individuals.

Little new was added to Hippocrates' views until the beginning of the present century. Some of his ideas have stood both the test of time and scientific scrutiny.

Among the early folklore is the simple observation that some people can feel the weather changing. Usually their complaints stem from a specific physical impairment. In many the sensitivity is induced by scar tissue.

Any extensive surgical procedure or amputation leaves scars. These are not only regenerated skin, but also usually involve some reconstituted muscle. This newly repaired cell structure is not in entirely uniform harmony with the older parts. In spite of all modern surgical skill it is still a patch job. It shows this by often reacting differently to environmental changes than the older tissue. This sets up an internal stress which results in pain. In people who have lost a limb, it often appears as if the missing part were aching. For this reason such pains have been given the label phantom pains. These pains are, however, not imaginary ills. They are very real. In fact, the ill can be noted even in well-healed simple scars from surgical incisions. In that case there is often no real sensation of pain, but rather a slight itching or tingling. Sometimes deformed skin tissue, as in the case of corns, can also become very painful.

Unfortunately, the last few wars have resulted in many major injuries, which provide a substantial group of persons who are victims of these pains. One thing is immediately clear from records kept on onset, intensity, and duration. The pains have a seasonal trend. They are more common in the months when the weather is wet and stormy than during the times of the year when dry and sunny conditions prevail. When we look at the day-to-day changes, we find again that the disturbed weather phases coincide with complaints about scar pains. High humidities invariably seem to accompany the weather miseries.

Some have speculated that the humidity has a direct effect on the skin itself. Skin is actually hygroscopic; in other words, it shows a slight expansion with rising humidity, and contracts a bit when the air becomes drier. In some cases, with outdoor exposure, this may trigger the pain, but it is very unlikely the cause in the vast majority of the victims who may stay indoors most of the time. Others ascribe pain to changes in the cell fluid or in the cell walls. Here again the electrical forces both in the environment and in the body may interact. Experimenters have been able to provoke in the laboratory scar pain in some individuals by alternating low-frequency electric fields.

By far the largest group of weather sufferers is among persons afflicted by ailments grouped under the general heading of rheumatism. This is more a collection of symptoms than a well-defined disease. The usual complaints are pains in muscles and joints. In bad cases they lead to deformities. Not all rheumatics have weather-related pains, but most of those who have chronic complaints react to the atmospheric pulsations. In some of the stricken, sharp reactions to falling temperatures occur. They can be helped by local or general application of dry heat, and it is not uncommon that their treatment will include a move to a warm, dry climate.

Rheumatoid arthritis, with involvement of the joints, is perhaps weather-bound as no other disease. It is clearly more prevalent in cold, moist, stormy climates than in warm, dry, calm ones. Locally, in the bad weather zones, people in poor housing are more likely to be affected than those living in well-heated places. The environmental conditions seem to act on the clear albuminous fluid that lubricates joints and tendons. Exactly through what mysterious pathways this comes about is not known. Yet when the weather changes to cold this fluid appears to thicken in the affected persons, and offers more resistance to motion. This higher resistance undoubtedly initiates the pains in joints. Also, most rheumatics have poor heat regulatory functions and have a much harder time adjusting to weather changes than normal persons.

Particularly revealing has been a series of experiments in which hospitalized arthritis patients were subjected to artificially induced atmospheric changes. For this purpose they were isolated in a room that was completely separated from the outside. All elements of the atmospheric environment could be separately controlled. This chamber is called a "climatron." The experimenters not only recorded the subjective reactions of the patients to the artificially induced changes, but also objectively measured their condition, such as swelling of joints. The results showed that weather pains are not a figment of the imagination, but supported by clinical findings. No clear relations emerged when only one element of the environment was varied. But when the barometric pressure was reduced and simultaneously the relative humidity raised, the typical arthritic complaints were provoked. These are also, of course, two of the typical changes that take place in weather phase 4.

Weather also aggravates another complex of diseases that are characterized by a single set of symptoms: asthma. These disorders culminate in breathing difficulties, which range from wheezing to almost complete suffocation. In some cases this is the reaction to a foreign substance, to which the body has developed an allergy. The atmosphere carries innumerable natural particles and many more are brought into the air as waste products of modern life. Some persons are extremely sensitive. They have allergic reactions to substances that are quite common in natural air. Many of these provocative particles are plant pollen. Trees, bushes, grasses, and weeds provide them in profusion in various seasons. The wind carries them shorter or longer distances, according to their size and aerodynamic characteristics. Some of them are really geared to glider flight. Even in a light wind fir pollen may carry several miles from their birth place. Fungus spores are particularly small and once brought into the air may stay until they are caught by hitting an obstacle or are washed out by rain.

Among the offensive pollen is, of course, ragweed, which causes misery to so many in the late summer and early autumn. Its pollen generally do not carry very far. Most of it is deposited within 50 meters of the producing plant. Unfortunately every plant produces millions of pollen and a few are carried to higher levels by the turbulent afternoon winds. And all too often their flight terminates in the nostril or mouth of a victim. There it produces the sneezing and wheezing of the allergic reaction.

Sometimes pollen are carried by the strong, gusty winds of an intense cold front for over 1000 kilometers. They can occasionally be traced back to a source region where the plant producing them happens to be ripe at the time. Should the particular pollen cause an allergic reaction in individuals, they may often wonder about their attack, especially if no local sources are present at the time. For seasonal "hay fever" sufferers, it usually takes another meteorological event to terminate their annual suffering. In case of ragweed, this is usually the first heavy freeze in autumn, which ends the life cycle of the plant.

The allergic asthma attacks stop as soon as the offending substances are removed, be they pollen, dust, or smog particles. In these cases the atmosphere only acts as a carrier. However, even when the air is clean, atmospheric events may trigger asthma attacks. These cases are much more difficult to explain. Generally they have a seasonal pattern. Early autumn is often the time when asthmatics of the non-allergic type have severe attacks. Their spasms are usually associated with the first invasions of cold air. The quiet, pleasant weather of late summer or early fall is suddenly brought to an abrupt end by an intense storm center. This usually brings first quite warm air, followed by a sharp cold front and rapidly falling temperatures. Suddenly hospital admissions for asthma attacks rise to two or three times their usual number. Later cold fronts during the winter may trigger further cases, but not as spectacularly. Figure 14 shows a particular incident that shows the trigger effect of falling temperatures on asthma attacks. The temperature curve shows the mean daily temperature value at Central Park in New York City.

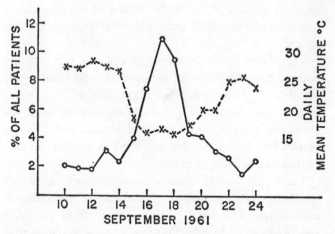

FIGURE 14. *Asthma attacks in early fall 1961 in a New York City hospital (solid line, left scale, in per cent of all patients treated), as related to temperature (dashed line, right scale).*

Note the rapid drop from September 14 to September 16, which is the effect of a strong cold front. At the same time the visits to the emergency clinic at Bellevue Hospital for asthmatic attacks rose spectacularly. These are shown by the solid line in per cent of the visits to the clinic for all types of ailments. After the weather warmed up again and became settled, the number of asthma patients treated at the hospital dropped back to the same level as prior to the onset of the cold front.

We are still ignorant of the exact way these weather conditions provoke the attack. It may or may not be a direct effect of the cold. One reason for assuming a direct effect is the fact that in many asthmatics, the capability to compensate for temperature changes is impaired. This deficiency is caused by changes in the delicate balance of the nervous system and certain hormonal secretions. For example, adrenal hormones are not produced in adequate quantity in some asthma sufferers. If they are supplied by medication at the right time, asthma attacks can be forestalled.

This shows for the first time clearly why it is very important for physicians and meteorologists to work closely together. A good forecast of impending weather events cannot only warn the physician of possible large increases in number and severity of attacks, but can also permit him to exercise his judgment in refining his preventive therapeutic measures.

There are some other chronic diseases in which weather plays a role, not as a cause, but by aggravating or ameliorating its course. Notable among these are heart and circulatory ailments. These are at present the greatest killers in the developed countries. Any set of statistics will show that these diseases have a very pronounced seasonal course. Usually, there is a peak of deaths in mid-winter, in January and February. In summer many fewer deaths occur. A more detailed analysis of the numbers of heart attacks and heart deaths reveals that the seasonal factor at work is the weather.

The weather influence on the person afflicted with heart disease appears to be both direct and indirect. Excluded here are the cases of heart attacks caused by over-exertion, such as shoveling snow. Presumably, any other kind of heavy work might have had the same result. There is, however, a definite effect of chilling which seems to cause adverse reactions. In part, cooling of the outer parts of the body may cause a greater stress on the heart. There is also a suspicion that breathing of very cold air places a very heavy burden on the heart-lung system. Persons who have insufficiencies of the coronary arteries, which supply the heart with blood, will get chest pains when walking against a cold wind.

The effect of cold weather on many people is to increase their blood pressure. This is an adverse circumstance for individuals whose circulatory system is already otherwise impaired. The effects of frontal passages, both warm and cold, with rapid, large temperature changes appear to have adverse, often fatal, effects on patients whose blood vessels suffer the usual deterioration with aging. Even if they are not directly exposed to outdoor conditions blood clots seem to occur with higher frequency on days with disturbed atmospheric conditions than on days with calm weather. Some scientists have again invoked electrical phenomena that are supposed to affect the thickness of the blood, but there is no proof of this hypothesis. Admittedly this remains a baffling puzzle.

Even though the winter peak of heart difficulties is well

established, one should not jump to the conclusion that all is safe for the heart patient in summer. Although warm, quiet weather is beneficial, there are heart conditions that have to be carefully watched even then. Among these is congestive heart disease, a condition in which a weakened heart muscle is unable to deliver the needed amounts of blood to the circulatory system and fails to empty the heart chambers of blood in its pumping motions. Persons who suffer from this ailment are very adversely affected by the combination of high temperature and humidity. Fortunately, they can be helped by air conditioning. In this case mechanical control of the environment can counteract the weather effect. In areas with a mild winter climate, summer is the season to be watched by the heart patient and his physician. Here too, prompt dissemination of pertinent weather information is highly desirable.

ELEVEN

Infections and Weather

Air carries a lot of organic material. Some, as noted above, causes allergic spells in sensitive persons. But among all this aeroplankton that may hit anyone are agents of disease, including spores of some fungi. One of them, which is a soil-dwelling fungus, deserves particular attention. It is a regular habitué of chicken coops. Once in a while a severe windstorm will stir it up from the soil and carry it along. Some meteorologists believe that it takes a tornado to dislodge them. This matters little to the poor victim who breathes it into the lungs. There this fungus causes a severe disease, called histoplasmosis, which in some respects resembles tuberculosis.

There are many other airborne diseases. Usually the infective agent, be it a bacterium or a virus, is expelled into the air by a human carrier through coughing or sneezing. These germs transfer from one person to another through close contacts in crowded rooms, public transportation vehicles, or through contaminated utensils. Their average transit time in the atmosphere is short and mostly indoors. Occasionally a few manage to get into the general air stream outdoors. There they probably have a very tough time to survive. They need moisture, but the relative humidity outdoors is often far from saturation. Under those conditions, the little droplets carrying the microorganisms dry out and the microbes, if they do not die, become at least dormant. The ultraviolet radiation from the sun, when it penetrates to the ground, is also a helpful agent in killing the germs. Still, a few survive. There is a strong suspicion that they become hitchhikers by attaching themselves to a larger dust particle. Some of these attract water vapor from the atmosphere even below saturation, and form the center of a small droplet. These drops may have a fairly long lifetime and act as sustainers of the microbes. Many of these small-scale processes have yet to be investigated in detail. They still offer many challenges to the research worker.

Actually only very few systematic observations, over longer intervals of time, have been made to ascertain the number of viable microbes in outdoor air. One set of measurements from Vienna, Austria, suggests a few general rules. There are fewer microorganisms in city parks than in heavily frequented streets. The number also decreases by about four-fifths from the center of the city to the suburbs. Right after rainfall, 40 per cent fewer were noted than in good weather, presumably a washout effect. And, finally, low temperatures brought about a decrease in the number of microorganisms.

Many disease-causing microorganisms are carried by insects. The weather affects them, too. Not only does the wind carry them hither and yon but they are subjected to heat, cold, and moisture conditions. A whole large science of biometeorological entomology deals with the effects of weather on insects. It is a field of great importance to mankind because insects also carry diseases to crop plants or act directly as destructive pests. In the context of our present story, it is only necessary to accept the fact that disease-carrying insects are ruled by weather conditions.

Finally, man's response to a germ is also conditioned by weather. His susceptibility seems to vary in an as yet unknown fashion from day to day. This confronts us with many ways in which weather can influence the outbreak of an infectious disease. It can act either on man or on the germ or on a carrier. Obviously, any two of these or all three may be affected by weather.

The following scheme shows the possible combinations. A + sign indicates a weather effect, a − sign, its absence.

MAN	PATHOGEN (Microbe-virus)	CARRIER
+	−	−
−	+	−
+	+	−
−	−	+
+	−	+
+	+	+

Thus in diseases in which all three are involved, we have six possible combinations for weather effects. If no carrier is involved there are only three combinations, which are marked off by the separating line. A typical disease of the type where all three are involved is malaria, where a bacterium is transmitted by a mosquito carrier to man. More about malaria will be related below. It is enough to say that only in very few cases have we been able to trace the weather effects through the whole chain of events. Unfortunately, it is only too reasonable to assume that the last combination with weather effects on man, the pathogen, and the carrier, is the common one.

Many infectious diseases are seasonal in incidence. In all such cases the weather is suspect as a contributing factor. Many of the respiratory infections have a peak in winter and occur less frequently in summer. Intestinal infections, in contrast, are more common in summer than in winter. The case for atmospheric influences is strengthened by the fact that the Northern and Southern Hemisphere show the disease peaks six months apart. In the tropics the respiratory ills show their greatest number during the rainy season.

In the past the number of diseases with seasonal peaks was considerably higher than now. Among them were many of the childhood diseases, including summer diarrhea of infants, diphtheria, whooping cough, poliomyelitis, and measles. These have been brought under control by improved sanitation and vaccines. Other diseases, such as typhoid fever and dysentery, which have been eliminated in countries with good sanitation, are still rampant elsewhere. They show a distinct summer peak. In these cases fresh foods, especially vegetables, act as important carriers. A warm, moist environment combines with unsanitary handling to create favorable conditions for survival and spreading of the microorganisms causing these diseases. This brief glance at the diseases, which hopefully will soon be completely controlled everywhere, may suffice. But there remains a formidable array of other uncontrolled infectious diseases which interact with weather.

The most important group here are the respiratory diseases. These are known to spread from person to person through fine droplets ejected by coughing or sneezing. The viruses are carried in these droplets, and move through the air just as the fine mist sprayed from an aerosol can containing insecticide. If the droplets are small enough they can float in turbulent air currents for a considerable length of time. Even after they sink to the floor or ground some will survive for some time, and may be stirred up again with the dust. The droplets are, of course, also affected by weather factors other than wind. Prominent among them is humidity. In air with low relative humidity, the droplets will evaporate and leave the microorganism without a base of fluid, which it needs for survival. High humidities prevent this evaporation and keep the droplet intact. In winter outdoor relative humidities are usually higher than in summer.

Another important factor is the ultraviolet radiation. These rays, which can so profoundly affect man's skin, are lethal to many microorganisms. In the higher latitudes, ultraviolet radiation is most intense at the earth's surface when the sun is high above the horizon. In many places summer is also the season with most sunshine. So, radiation is apt to combine with the low humidities to eliminate a lot of pathogens. In winter ultraviolet radiation is usually blocked by clouds or by the thick layers of polluted air the sun's rays have to penetrate. This and the high humidity favor survival of germ-carrying droplets.

On the other hand, very low temperatures, which characterize winter in some regions, are unfavorable to the droplets. Even very small droplets will freeze at about $-15°$ C, and thus cease to be a swimming pool for microbes. In fact, respiratory diseases are rare in the polar regions. If they occur they are usually imported by recent arrivals and rapidly disappear when those afflicted have recovered. Thus outdoors, a damp, cloudy, cool climate favors the survival of respiratory germs. The Atlantic

shores of northwestern Europe in winter fill these speci-
fications particularly well. Incidence of colds, bronchitis,
and influenza is high there.

Yet the outdoor conditions hardly tell the whole story.
We must take indoor conditions into account too. Some
of these are more sociological than meteorological. People
are crowded into public conveyances with windows
closed; ventilation in homes is poor; with the start of
cooler weather the school, concert, and theater seasons
start. All these circumstances bring people into close contact in an environment of stagnant air. Obviously a few
droplets from a sneeze will quickly find an unwary victim. Even the low humidities usually encountered indoors
during the heating season, although shortening the lifetime of droplets, may not be a really favorable circumstance. It may just tend to keep the droplets very small,
and therefore in suspension. With high humidities the
droplets may stay large or even grow. This will favor
settling to the floor. Settlement actually is much easier
indoors where there is little air flow and generally no
turbulence. This is in contrast to outdoor conditions
where even a slight breeze will have turbulent motion
that will help in keeping suspensions afloat. We know
nothing yet about other weather influences on pathogens.
Some scientists have speculated that they might be more
virulent at one time than another, but there is little
known to support this idea.

Now for a look at the victim of the microbic attack.
There is some evidence that weather affects man's recep-
tiveness. His resistance is a problem in connection with
seasonal infectious diseases. Does it have an annual swing
too? It is quite clear that an individual is physiologically

not the same spring, summer, fall, and winter. Some of the differences are simply adaptations to the changes in the environment. Others are produced by shifts in the balance of hormones. In addition, there are the often radically different modes of life between winter and summer. The question is: Do these shifts lower resistance at some times?

The evidence is quite flimsy. Some suspect a direct humidity effect on the mucous membranes of nose and throat. Remember here that cold air has of necessity a low moisture content. When it is warmed during inhalation and moistened to saturation in exhaling, it has a tendency to dry out the moist surfaces of the respiratory tract. This is assumed to permit easier penetration of microbes into the tissue and blood stream. There they meet another defense of the body, the white blood corpuscles, or leucocytes, which will try to overcome the invaders. Their success may depend on their number, and this number of leucocytes seems to undergo a seasonal and a day-to-day variation. The former is such that leucocytes reach a maximum in early spring and a minimum in late summer. This is what one would expect in any army during wartime; a steady build-up during the fighting, with the greatest number at the end of the threat and a reduction during periods when fewer invasions threaten. We may speculate here that in autumn, when the infection season starts, this part of the body's defenses is not yet in the "ready" state. Thus the first cold of the season catches us unprepared. The leucocytes also seem to rise in number during frontal weather conditions, and fall when the weather is quiet. Is the body here alerting its defenses to counteract invasions of infectants?

It might well be an age-old reaction to natural forces. It is also well warranted. Many studies have shown that upper respiratory infections have a large weather element associated with them. By statistical analysis one can eliminate the wide seasonal swing, and then study the effect of individual temperature fluctuations separately. This shows, particularly in older people, falling temperatures and upper respiratory infections to be associated. The analysis is quite complicated because there is usually an incubation period of several days after the infection before disease symptoms appear. Also, many persons do not seek medical attention or go to clinics, so that only the more severe cases enter into the records. Generally, statistics of absenteeism in industry give a better measure of the quick fluctuations in upper respiratory infections. These statistics also support the contention that sudden weather changes, especially cold fronts, contribute to the number of victims of colds and flu.

There is still a fierce controversy over whether or not chilling has anything to do with catching cold. Obviously, chilling, without exposure to the virus, cannot provoke the disease. Some experiments have been carried out with volunteers, who were exposed to a cold atmosphere and had their feet immersed in cold water. Yet there was no greater number of colds among them than others who had not been chilled. Such experiments often do not involve the age groups which are most afflicted. Neither have they so far duplicated all the subtle simultaneous changes that take place in the atmosphere when a cold front passes. So the question remains open and there is only the evidence that sickness statistics offer. The results of one such study are shown in Figure 15. These data

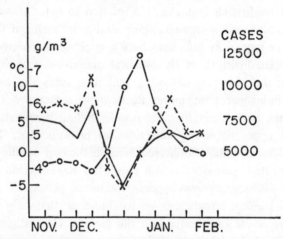

FIGURE 15. *Weekly cases of respiratory diseases (long dashes, right scale) in relation to temperature (solid line, outer scale on left) and absolute humidity (short dashes, inner scale on left).*

show weekly information. This perhaps hides some details, but at the same time smooths out the uncertainties about incubation time and reporting. How do we explain the fact that after a major drop in temperature and vapor pressure the respiratory ailments rose? Coincidence? Perhaps, but similar combinations occur year after year. The development of protective vaccines may well occur before we have an answer to these puzzling questions. Thereafter, nobody may care whether our last cold came because we had not worn galoshes and mufflers.

It remains to be seen what happens when a carrier gets in the middle between pathogens and their unwilling hosts. This is very common in the so-called tropical diseases. In many of these, insects are the carriers. It took a lot of scientific detective work to discover this. The best-known cases relate to yellow fever and malaria. Both of them are mosquito-borne.

There is no evidence that in these diseases the influence of weather on man plays a role. Nor is it known whether or not Plasmodium, the malaria parasite, for example, is directly affected by weather. In any event, these would undoubtedly be quite minor influences compared to the influence of weather on the mosquito. Although there are a large number of mosquito varieties with different responses to the environment, they have some common reactions. Our attention shall be focused here on Anopheles, the malaria carrier. The larva are laid into stagnant waters, from which they emerge upon maturation as those baneful buzzers. This fixes immediately a weather dimension because the development of stagnant water pools or puddles requires a minimum amount of rainfall. We can set it at about 1000 millimeters per year. The mosquito also needs warm temperatures. They do not seem to breed when the mean monthly temperature is below 16° C. Monthly mean temperatures between 22° and 25° C suit them best. Low wind speeds seem to be another of their favorite conditions. Their nocturnal habits are favored by the calmest hours of the day. Actually they do not migrate far from the birthplace, generally not more than a kilometer.

With this knowledge it is fairly simple to outline areas on the globe where malaria could be endemic. On climatic charts, the 15° C isotherm, the 1000 millimeter isohyet, and areas of low wind speed are readily located. A composite of these is shown on a world map in Figure 16. The hatched areas are those where malaria mosquitoes could readily survive and the stippled regions show localities where they might offer temporary threats, especially

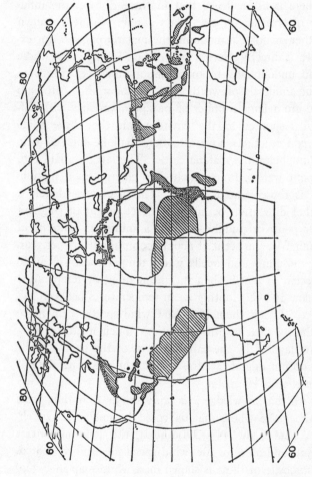

FIGURE 16. Areas in world where climatic conditions are permanently favorable (hatched areas) or temporarily favorable for malaria.

near poorly drained swamps. Modern hygienic measures attempt to fight the mosquito by dichlorodiphenyltri-chloroethane, a tongue-twisting name usually abbreviated as DDT. Although some progress has been made in this campaign, together with screens and mosquito netting, still large population groups suffer from malaria, which undermines their strength.

Similar climatic considerations permit us to outline the limits of other insect-carried diseases. Yellow fever, which has already been mentioned, is also being countered by an effective vaccine. Sleeping sickness, transmitted by the climate-bound tsetse fly, carrying trypanosoma, falls into the same group of ailments. Another tropical disease governed by temperature and rainfall conditions is yaws. It is caused by one of the Treponema spirochetes. It is not certain that an insect is involved in the transmittal of the disease, but the microbe seems to get along just fine in warm, moist soil. It causes direct infection through insect bites and scratches and particularly endangers people with bare feet. The distribution of the disease is entirely within the 24° C annual isotherm, but the rainfall required is only seasonal and need not reach the amount favoring mosquitoes.

If all this sounds complex, nature has conjured up still more complicated tricks. An example is plague. Two hosts are involved: one, a rodent, usually a rat; the other, a flea. The rat can generally manage to survive in most weather zones, except the polar regions, but the flea requires high temperatures and humidities for survival. However, potentially large areas on earth would fill their climatic needs. A final example of climatically controlled diseases are those of the hookworm type. These worms, too, have multiple hosts in their life cycle, such as snails and water fowl. Their survival depends on large amounts of rainfall and periodic flooding. These are most common in the humid tropics, where the hookworm diseases are rampant.

TWELVE

Climate, Acclimatization, and Therapy

Weather and climate not only play a role in causing disease but also relate to the healing process. They can become adjuvants to recovery and are deliberately used by physicians in the care of the ill, or as an aid in the care of the aged. But occasionally they become inadvertently involved when therapy centers around drugs.

Nearly all drugs are tested in animals for effects and potency. Many of them are biologically standardized prior to use by animal reactions. For potentially dangerous drugs this is common practice. The tests assure the safety of various batches manufactured by the same plant and guard against differences in the products of various manufacturers.

One of these drugs, with widespread use in human heart disease, is digitalis. It was found years ago in testing this drug that equal doses from the same batch showed day-to-day differences in potency. Usually the tests are performed with cats, but other test animals show the same responses. The strange results of the tests showed that the lethal doses of digitalis were larger on some days, smaller on others. The fluctuations in response seemed to follow the weather conditions. A series of experiments established that this drug is more toxic during storms than in quiet conditions. In weather phases 4 and 5 the effects are more notable. It was also shown that the digitalis response increased with body temperature. For each degree centigrade temperature rise, the toxic effect rose 10 to 15 per cent. Except in fevers the body temperature will increase only in extreme heat so that this direct temperature effect will arise only rarely in man from atmospheric influences. But there seem to be other weather-steered reactions. One has been attributed to pressure. The toxicity rises with elevation. Test animals in the mountains were killed by smaller doses than control animals at sea level.

At higher elevations this may be an effect of the reduced oxygen pressure. It is well known that, for example, alcohol has a greater intoxicating effect at the reduced oxygen level of high mountains than at sea level. Evidently, at higher altitudes the alcohol that circulates with the blood is "burned" at a slower rate.

Yet the direct temperature and pressure effects are not a complete answer to the weather-related drug reactions. Some believe that weather acts like a stress on the human body. Under this hypothesis a stress affects the permeability of various membranes in the body. This means that in some conditions a drug can enter the blood stream more readily than under others. This again confronts us with the fact that the beginning of the effect and the end, but very little about the mechanism that links the two, are known. Weather sets the reaction into motion and the end result is higher or lower toxicity of drugs. But the precise biological pathway of this effect is still a mystery.

These effects are particularly important in sedatives and narcotics. Stress causes the same doses of morphine or chloroform to cause narcosis faster than in the absence of stress. Tests in mice have shown, for example, that during frontal weather disturbances the lethal doses of morphine sulfate are smaller than in quiet weather. But the stress effect of weather is not one-sided. In some drugs the effects are increased; in others, decreased. In some drugs inducing sleep, the effectiveness is reduced at high temperatures.

Insulin, widely used to combat diabetes, is also weather-dependent in its action. Exposure to cold apparently slows the response to this drug.

Quite important also are the actions of drugs that may interfere with the heat-regulatory mechanism of the body. An example of this type is atropine. It is well known for its action in dilating pupils, but it is also used for relieving spasms. In that form it is taken internally. One of its side effects is the inhibition of sweating. In persons exposed to a hot environment this will prevent evaporative cooling from the skin and consequent rise of body temperatures to the point of heat stroke.

In hot weather conditions considerable precaution is also indicated for the use of certain diuretics. This is particularly important for persons not acclimated to a hot environment. They usually have low urine secretion and high loss of body sodium. Diuretics under such conditions may cause excessive loss of sodium which in turn may lead to circulatory failure.

It cannot be our task to review here other individual drug effects in their relation to weather. The important fact to stress is that such effects exist. Patients in hospitals and under steady care of physicians will have their medications properly adjusted. Some people, however, receive prescriptions and take drugs in constant quantities without much supervision. They may be the ones most exposed to the weather-induced side effects of drugs. People moving in the course of business or during their vacations rapidly from one climate to another are most endangered by dosages of drugs improper for the different climatic environments. Travelers are therefore well-advised to check with their physicians about their medications before going through drastic changes in atmospheric conditions. In today's jet travel age a shift from the arctic to the tropics is just a matter of hours. This may well have considerable consequences on the effect of our favorite pills.

Yet not only the drug-taker, but also the healthy individual can be markedly influenced by a rapid change in climate. Individual reactions to a rapid change vary widely. Some persons will adapt quickly to a new environment, others slowly. Much depends on age, body build, and—often most important—common sense. The latter implies that, to make a smooth transition from one climate to another, people have to change their habits of clothing and eating to suit the new environment. Changing one's clothing is generally easier done than changing food habits. Tourists will readily slip into the right togs to stay comfortable. Even armies have learned to issue clothing adapted to the climate. This was not always the case and much misery has come to soldiers because regulations issued for a moderate climate were applied to men fighting in an arctic or tropical environment.

Fluid intake is usually quickly adjusted to the environmental demands. Thirst is such an insistent and unpleasant sensation that it is not likely to be ignored. For the caloric intake adjustments are much slower. The whole process of acclimatization may take weeks or months even in healthy young people. Many travelers move so rapidly through a climatic zone that they never have time to adapt. Air conditioning and heating of airplanes, vessels, and hotels have made climatic extremes more accessible and acceptable to the traveler than in prior ages. These conveniences may not be available to more permanent migrants. In their case, for optimal adaptation, major adjustments in the endocrine system are needed.

Some persons can never fully adapt to major climatic changes. Some may adjust only in one direction; they may tolerate a move from moderate to warm, but not to colder, or vice versa. A major climatic change may impair, temporarily or permanently, performance of physical labor, a problem that often arises in connection with sports contests. Athletes who want to compete in a climate other than that of their home base, often train for weeks in order to achieve top form. But although many of these reactions are steered by hormonal balances, it is now known that motivation also plays an important role. Racial characteristics seem to be of subordinate importance. Children born to immigrants from another climatic zone seem to be as readily adapted to the new environment as descendants of indigenous strains.

Acclimatization works both ways. A person who came to the tropics from a moderate climate, after several years, requires again an acclimatization period when he returns to his home. In some persons, who retire after many years of service in a tropical climate to a cool region, the required elasticity has been impaired by age and they may never completely adapt again, even though they are in their native atmospheric environment.

Physiological (not chronological) age seems to have a great deal to do with reactions to changes in weather and climate. Most young and healthy persons can adapt to the environmental changes readily and rapidly. In older persons, even without physical impairment, the adjustments take a longer time. In case of rapid weather changes this can mean that the aged body always lags behind the events in its attempts to keep all factors in equilibrium. Thus, a person may always be a bit out of balance and not feel well even though he has no specific disease. This has led to the gerontological recommendation that older persons are better off in comfortable climates with steady weather conditions than in climates that cause large physiological demands and where weather changes are severe and rapid.

It is easier to define these retirement climates meteorologically than to find them in nature. The desired circumstances call for moderate temperatures and humidities, and low storm frequency. Such conditions are most common at the poleward fringes of the subtropical high-pressure cells of the general atmospheric circulation. In the Northern Hemisphere, these cells, which are very persistent, have their northern fringes around 25 to 35° latitude. Meteorologists have given them names, such as the Bermuda High or the Pacific High. Weather under their control is usually very stable but not necessarily always comfortable. Maximum comfort in their realm is often right close to a seashore or on an island where temperatures are even but humidities can at times be high. In the United States, the coastal area around San Diego, California, comes closest to the postulated conditions. Some coastal areas of the Mediterranean, coastal southern Chile, the east coast of southern Australia are among the areas that approximate similar, favorable climates. Many countries, of course, have no climatic zones that are permanently in such favorable conditions. A good many do have seasonally an atmospheric environment that fills the criteria. This often leads to a migratory existence as a good way to cope with the atmospheric vagaries. In northern Europe it has long been a practice to have older people, if possible, spend the winter months in Spain, Italy, Greece, the Crimea, or North Africa. In the eastern United States the trek to Florida and the Caribbean Islands in winter is a comparable scheme.

How much an individual can even out the climatic conditions surrounding him is best illustrated by an example. In this case it is assumed that the person spends the winter in Florida and the summer in New Jersey at coastal localities. The temperatures for two places in these areas are shown in Figure 17. The winter cold, which is associated with many storms and fronts, is in the northern locality and fairly high temperatures prevail in the south during the summer. Both of these extremes can be avoided by a northward move in late April and a southward move in October. Other similar pairs of localities could, of course, be found.

Climate is not only a health aid for the old but has been used as health restorer since the earliest ages in medicine. For a while in the last century it was considered to be one of the most potent weapons in the physician's arsenal. At one time it was the only remedy in the fight against tuberculosis. In the last half-century, chemotherapy has replaced climatic therapy in this and most other diseases. Yet climate remains as an important aid to physical medicine. In some countries, it plays a much greater role than in the United States.

For some ailments a change in atmospheric environment is still the only escape from ill-health. This is the case for many air-borne allergies, but also for those persons who are subject to allergic reactions from cold. Cli-

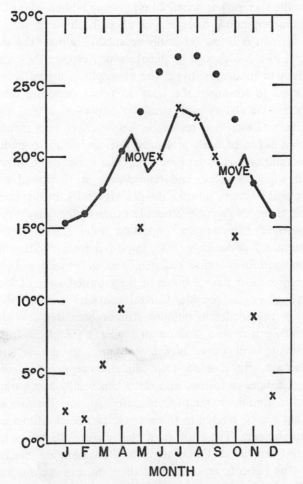

FIGURE 17. *Mean monthly temperatures at Daytona Beach, Florida (circles), and Atlantic City (crosses). Line connects months, showing a relatively even climatic exposure for a person moving south in the fall and north in the spring.*

matic therapy has beneficial effects on children who suffer
from respiratory diseases and whose health and state of
nutrition is below medically acceptable norms. The general idea is a change in climate which exposes these children to moderate stress. The atmospheric action is supposed to stimulate the body to better muscular tonus,
better circulatory and glandular responses, better appe-
tite, and increased resistance. To be effective, a climatic
cure has to be fairly prolonged because initially a gradual
adaptation has to be promoted. The atmospheric stresses
of wind, sunshine, and temperature are increased over
several weeks to provoke the desired results. In practically
all these therapeutic attempts other factors play simultaneously an equally important role. These include
improved personal hygiene, improved food, sports, com-
panionship of other children, and in general a better
environment than a home in the crowded areas of modern cities can provide. Careful controls, however, show
that the moderate outdoor stresses presented by wind
and weather on a seashore or moderately high mountain
climate will cause higher resistance to disease even
months after a return from climatotherapy.

Changes in climate also often beneficially affect adults
with chronic respiratory ailments. Among the diseases
that can be treated in this way are bronchial asthma and
emphysema. These are both characterized by heavy mucous secretions. Often respiratory passages are blocked.
Many patients suffering from these diseases are much re-
lieved by a change in climate. The conditions to be suitable, as auxiliaries to other therapy, are pure air and low
humidities. Deserts often have too much dust present to
give an ideal environment, but the lee side of high mountains, especially at moderate elevations above the valley

floor, often offer climatic conditions that are particularly favorable for alleviating these ailments. In the United States the east slopes of the Sierra Nevada and the Rocky Mountains have many locations that have generally low humidities and yet offer some moderate climatic stresses that seem to benefit patients with asthma.

For persons with heart and circulatory disturbances, the same mild and stable conditions that seem to benefit older persons are advocated. Extreme wind chill and excessive heat are particularly detrimental. The heart patient also seems to have more difficulties in coping with rapid weather changes than a normal person. It is not yet clear whether or not complete seclusion from the outdoor atmosphere through air conditioning is as effective as a move to a quiescent climate. Air conditioning is, however, definitely beneficial.

Climatotherapy in recent years has also included the creation of entirely artificial atmospheric environments. Some of these occur in nature only rarely; others, never. These conditions are created in chambers where the composition of the air can be completely controlled and all environmental factors be regulated. These chambers serve primarily for the treatment of noninfectious respiratory difficulties. They also alleviate breathing difficulties that remain after the infectious phases of respiratory ailments are over.

Most interesting and least explained among these treatments are the beneficial effects of an atmosphere that carries large amounts of small negative ions in it. These are electrically charged air particles in very high concentration, generally about 3000 to 7000 per cubic centimeter. This number is about ten times higher than that ever observed in nature. These ions beneficially affect asthma sufferers. The value of positive ions has not been completely established.

Other treatments in chambers include both increased and decreased air pressure. Decreased pressure is less often used than increased pressure. Beneficial effects have, however, been noted in pressure decreased down to about 60 per cent of sea-level pressure for patients with bronchitis and whooping cough. These exposures are usually short and rarely exceed thirty minutes. Increased pressures, up to 50 per cent higher than normal atmospheric pressure at sea level, have also been used. Here again the beneficiaries have been persons with emphysema, chronic bronchitis, and aftereffects of pleurisy.

Ultraviolet radiation has also been used therapeutically, primarily for skin disorders. The artificial sources of these rays generally create wave lengths that are much shorter than those found in the solar spectrum at the earth's surface. Exposures to these rays have to be, of course, very brief to prevent tissue damage.

Postscript

This ends a brief account on weather and health. The reader must have sensed now that much remains to be discovered in this field. The explanations of how the weather interferes with the delicate balances of man's body mechanism are yet very weak. There are too many missing links to satisfy scientists. These gaps in knowledge can only be filled by patient collaboration between physicians and meteorologists. The dialogue between these two professions has been increasing in recent years. The day is not far off when the daily weather forecast will contain information on chill factors, comfort conditions, expected concentration of pollutants, and other biometeorological information. Similarly, physicians will advise their patients on precautionary measures to take under certain weather conditions. They may even heed the weather when they fill in that always-handy prescription blank. But the most important thing yet to be accomplished, is to teach modern man how to live in harmony with his environment for optimal health.

Appendix

CONVERSIONS

Temperature: Degrees Centigrade (or Celsius) to Degrees Fahrenheit

$$°F = \tfrac{9}{5}° \ C + 32 \qquad 1° \ C = 1.8° \ F$$

Some equivalents:

°C	−40	−30	−20	−10	0	10	20	30	40
°F	−40	−22	−4	+14	32	50	68	86	104

Atmospheric pressure:

1000 millibars = 750.06 millimeters of mercury (Torr) = 29.53 inches of mercury. Average sea level pressure 1013.3 mb = 760.0 mm Hg = 29.92 in. Hg.

Vapor pressure:

1 millibar = 0.75 millimeters of mercury (Torr)
  = 0.0295 inches of mercury.

Some equivalents:

mb	5	10	15	20	25	30	35	40
mm Hg	3.75	7.50	11.25	15.0	18.75	22.50	26.25	30.00
in Hg	0.148	0.295	0.443	0.591	0.738	0.886	1.034	1.181

Radiation:

1 langley/minute = 1 cal/cm^2 min = 14.33 watts/cm^2
  = $4,521 \times 10^{-3}$ BTU/ft^2 hr.
1 calorie = 4.184 joule.

Wind speed:

1 m/sec = 1.94 knots = 2.24 mi/hr.

Some equivalents:

m/sec	1	3	5	10	20
km/hr	3.6	10.8	18.0	36.0	72.0
mi/hr	2.2	6.7	11.2	22.4	44.7
knots	1.9	5.8	9.7	19.4	38.9
ft/min	197	591	984	1969	3937

Length (distance, height):

1 cm = 0.3937 inches

1 m = 3.2808 feet

1 μ (micron) = 1/100,000 cm
$ = 1/1,000,000$ m
$ $ or micrometer

1 mμ (millimicron) = 1/1000 μ
$ $ also called nanometer

Area:

1 sq cm (cm^2) = 0.155 sq in (in^2)

1 sq m (m^2) = 10.76 sq ft (ft^2)

1 sq km (km^2) = 0.386 sq mi (mi^2)

Volume:

1 cu cm (cm^3) = 0.061 cu in (in^3)

1 cu m (m^3) = 35.31 cu ft (ft^3)

1 liter = 33.815 fl oz = 0.264 gal (U.S.)

Mass:

1 kg = 35.27 oz = 2.205 lb

1 metric ton (t) = 2204.6 lb = 1.1023 short tons

Heat:

1 kilocalorie (kcal) = 1000 gram calories (cal)
$ = 3.968$ BTU (British thermal units)

Additional Reading

Bresler, Jack B. *Human Ecology*. Reading, Mass.: Addison-Wesley, 1966. 472 pp.

Folk, G. Edgar, Jr. *Introduction to Environmental Physiology*. Philadelphia: Lea & Febiger, 1966. 308 pp.

Griffiths, John F. *Applied Climatology*. London: Oxford University Press, 1966. 118 pp.

Licht, Sidney (ed.) *Medical Climatology*. Physical Medicine Library, Vol. 8. New Haven, Conn.: Elizabeth Licht, 1964. 753 pp.

Sargent, Frederick, II, and Tromp, Solco W. (eds.) *A Survey of Human Biometeorology*. Technical Note No. 65. Geneva: World Meteorological Organization, 1964. 113 pp.

Tromp, S. W., *et al*. *Medical Biometeorology*. Amsterdam: Elsevier Publishing Co., 1963. 991 pp.

Glossary

ACCLIMATIZATION   the gradual accommodation of physiological processes to a new climatic environment.

ADRENAL   pertaining to a small ductless gland located adjacent to the upper part of the kidney, secreting a hormone raising the blood pressure.

AIR MASS   a large body of air that acquires through surface contact uniform characteristics of temperature and humidity. Primary air masses are labeled for their origin—tropical, polar, maritime, or continental.

ALLERGY   an excessive sensitivity to substances breathed or ingested or to other environmental conditions, causing reactions such as hives, sneezing, or asthma.

CIRCADIAN   a rhythm during the course of a calendar day, usually applied to physiological phenomena.

CLIMATE   the collective state of weather at a place or over an area, expressed by the statistical properties of various weather elements. The average of temperature and the total of precipitation for a given time interval are often chosen as characteristics of climate.

CONTINENTAL CLIMATE   a climate characterized by wide range in temperature between the warmest and coldest month, apt to have a dry, sunny regime, often with a pronounced rainy season.

CONVECTION   mass movement of air, often specifically applied to vertical movements induced by radiative heating of a surface, which causes buoyancy in the overlying air.

CYCLONE   a center of low air pressure around which winds circulate in counterclockwise direction in the Northern Hemisphere (clockwise in the Southern). Examples in ascending order of size are: tornado, hurricane or typhoon, extratropical storm.

ENDOCRINE   pertaining to body glands producing internal secretions which, carried by blood or lymph, control or regulate various organs or functions.

FRONT   a boundary or boundary zone between atmospheric air

masses, usually labeled, for the temperature of the actively advancing air, *cold* or *warm* front.

GANGRENE, GANGRENOUS decay of body tissue usually because of obstruction or loss of blood supply to an injured area of the body.

HIGH-PRESSURE AREA OR SYSTEM a center or region of high air pressure, inducing in the Northern Hemisphere a clockwise outflowing circulation (counterclockwise in the Southern); an anticyclone.

HOMEOTHERMY, HOMEOTHERMAL the property of warm-blooded beings to keep approximately constant body temperature.

HORMONE a substance formed by and circulated from one of the glands of internal secretion, generally causing powerful stimuli or regulation of body functions.

HUMIDITY a measure of the content of water in gaseous form in the air; expressed as absolute humidity in terms of grams of water per cubic meter of air or partial pressure of the vapor (see VAPOR PRESSURE); as relative humidity, expressed as water vapor actually present in per cent of that maximally possible at the prevailing temperature.

INCUBATION TIME the interval between infection with a disease and the appearance of symptoms.

INFRARED part of the spectrum of electromagnetic waves beyond the visible red but of shorter wave length than radio waves; infrared waves are often designated less precisely as heat waves.

INVERSION in meteorology, refers to a vertical temperature distribution where warmer air is lying over cold air; specifically a ground inversion where air temperature is lowest near the ground and increases with elevation; a stable stratification of air in which vertical motion is inhibited.

ION in meteorology, electrically charged air molecules or particles floating in air.

ISOHYET a line along which the amounts of rainfall are equal.

ISOTHERM line along which the air temperature has the same value.

LOW-PRESSURE AREA OR SYSTEM a center or region having low barometric pressure, usually applied to a large cyclone (see that term).

MARITIME CLIMATE a climate characterized by a small range between the warmest and coldest month of the year; diurnal variations of temperature are also small; often cloudy and marked by an even distribution of precipitation throughout the year; humidities are usually high.

METABOLISM a process characteristic of living beings in which food substances are transformed into tissue with release of energy and waste.

NECROTIC   pertaining to dead tissue in the body.

OCCLUSION, OCCLUDED CYCLONE   a storm system in which the cold air has completely overtaken the warm air and lifted it off the ground.

PATHOGEN   a microorganism or virus causing disease.

PATHOLOGY   the branch of medicine dealing with disease, especially the morphological and functional changes caused by disease.

PHOTOCHEMICAL   narrowly pertaining to chemical actions induced by light, but more generally including changes caused by other wave lengths, such as ultraviolet; attributed to photons, which are quanta of energy.

PHYSIOLOGY   the branch of biology dealing with the functions of and processes in living organisms, usually referring to normal conditions.

RADIATION   transmission of energy by electromagnetic waves through space or gaseous medium.

RELATIVE HUMIDITY   the amount of water vapor present in a given space in per cent of what this space could hold maximally at the prevailing temperature (see also HUMIDITY).

SPECTRUM   the range or dispersal of the various wave lengths from a light or energy source.

SYSTOLIC BLOOD PRESSURE   arterial pressure of the blood at the time of heart contraction when blood is forced from the heart chambers into the circulatory system.

TOXICITY   the state or level of poison or drug action.

ULTRAVIOLET   part of the electromagnetic wave spectrum shorter than the visible violet light but longer than x rays; a photochemically active radiation.

VAPOR PRESSURE   the partial pressure of water vapor mixed with air.

VASOCONSTRICTION   constriction of a blood vessel or blood vessels.

Index